The Human Brain during the Third Trimester 310- to 350-mm Crown-Rump Lengths

This thirteenth of 15 short atlases reimagines the classic 5 volume *Atlas of Human Central Nervous System Development*. This volume presents serial sections from specimens between 310 mm and 350 mm with detailed annotations. An introduction summarizes human CNS developmental highlights around 9 months of gestation. The Glossary (available separately) gives definitions for all the terms used in this volume and all the others in the *Atlas*.

Key Features

- Classic anatomical atlases

- Detailed labeling of structures in the developing brain offers updated terminology and the identification of unique developmental features, such as germinal matrices of specific neuronal populations and migratory streams of young neurons

- Appeals to neuroanatomists, developmental biologists, and clinical practitioners

- A valuable reference work on brain development that will be relevant for decades

ATLAS OF
HUMAN CENTRAL NERVOUS SYSTEM DEVELOPMENT
Series

Volume 1: The Human Brain during the First Trimester 3.5- to 4.5-mm Crown-Rump Lengths

Volume 2: The Human Brain during the First Trimester 6.3- to 10.5-mm Crown-Rump Lengths

Volume 3: The Human Brain during the First Trimester 15- to 18-mm Crown-Rump Lengths

Volume 4: The Human Brain during the First Trimester 21- to 23-mm Crown-Rump Lengths

Volume 5: The Human Brain during the First Trimester 31- to 33-mm Crown-Rump Lengths

Volume 6: The Human Brain during the First Trimester 40- to 42-mm Crown-Rump Lengths

Volume 7: The Human Brain during the First Trimester 57- to 60-mm Crown-Rump Lengths

Volume 8: The Human Brain during the Second Trimester 96- to 150-mm Crown-Rump Lengths

Volume 9: The Human Brain during the Second Trimester 160- to 170-mm Crown-Rump Lengths

Volume 10: The Human Brain during the Second Trimester 190- to 210-mm Crown-Rump Lengths

Volume 11: The Human Brain during the Third Trimester 225- to 235-mm Crown-Rump Lengths

Volume 12: The Human Brain during the Third Trimester 260- to 270-mm Crown-Rump Lengths

Volume 13: The Human Brain during the Third Trimester 310- to 350-mm Crown-Rump Lengths

Volume 14: The Spinal Cord during the First Trimester

Volume 15: The Spinal Cord during the Second and Third Trimesters and the Early Postnatal Period

The Human Brain during the Third Trimester
310- to 350-mm Crown-Rump Lengths

Atlas of Human Central Nervous System Development, Volume 13

Shirley A. Bayer
Joseph Altman

CRC Press
Taylor & Francis Group
Boca Raton London New York

CRC Press is an imprint of the
Taylor & Francis Group, an **informa** business

Designed cover: Shirley A. Bayer and Joseph Altman

First edition published 2024
by CRC Press
2385 NW Executive Center Drive, Suite 320, Boca Raton FL 33431

and by CRC Press
4 Park Square, Milton Park, Abingdon, Oxon, OX14 4RN

CRC Press is an imprint of Taylor & Francis Group, LLC

LCCN no. 2022008216

ISBN: 978-1-032-22885-3 (hbk)
ISBN: 978-1-032-22880-8 (pbk)
ISBN: 978-1-003-27464-3 (ebk)

DOI: 10.1201/9781003274643

Typeset in Times Roman
by KnowledgeWorks Global Ltd.

Publisher's note: This book has been prepared from camera-ready copy provided by the authors.
Access the Support Material: www.routledge.com/9781032228853

CONTENTS

ACKNOWLEDGMENTS

We thank the late Dr. William DeMyer, pediatric neurologist at Indiana University Medical Center, for access to his personal library on human CNS development. We also thank the staff of the National Museum of Health and Medicine that was at the Armed Forces Institute of Pathology, Walter Reed Hospital, Washington, D.C. when we collected data in 1995 and 1996: Dr. Adrianne Noe, Director; Archibald J. Fobbs, Curator of the Yakovlev Collection; Elizabeth C. Lockett; and William Discher. We are most grateful to the late Dr. James M. Petras at the Walter Reed Institute of Research who made his darkroom facilities available so that we could develop all the photomicrographs on location rather than in our laboratory in Indiana. Finally, we thank Chuck Crumly, Neha Bhatt, Kara Roberts, Michele Dimont, and Rebecca Condit for expert help during production of the manuscript.

AUTHORS

Shirley A. Bayer received her PhD from Purdue University in 1974 and spent most of her scientific career working with Joseph Altman. She was a professor of biology at Indiana-Purdue University in Indianapolis for several years, where she taught courses in human anatomy and developmental neurobiology while continuing to do research in brain development. Her lengthy publication record of dozens of peer-reviewed, scientific journal articles extends back to the mid 1970s. She has co-authored several books and many articles with her late spouse, Joseph Altman. It was her research (published in *Science* in 1982) that proved that new neurons are added to granule cells in the dentate gyrus during adult life, a unique neuronal population that grows. That paper stimulated interest in the dormant field of adult neurogenesis.

Joseph Altman, now deceased, was born in Hungary and migrated with his family via Germany and Australia to the US. In New York, he became a graduate student in psychology in the laboratory of Hans-Lukas Teuber, earning a PhD in 1959 from New York University. He was a postdoctoral fellow at Columbia University, and later joined the faculty at the Massachusetts Institute of Technology. In 1968, he accepted a position as a professor of biology at Purdue University. During his career, he collaborated closely with Shirley A. Bayer. From the early 1960s-2016, he published many articles in peer-reviewed journals, books, monographs, and free online books that emphasized developmental processes in brain anatomy and function. His most important discovery was adult neurogenesis, the creation of new neurons in the adult brain. This discovery was made in the early 1960s while he was based at MIT, but was largely ignored in favor of the prevailing dogma that neurogenesis is limited to prenatal development. After Dr. Bayer's paper proved new neurons are added to granule cells in the hippocampus, Dr. Altman's monumental discovery became more accepted. During the 1990s, new researchers "rediscovered" and confirmed his original finding. Adult neurogenesis has recently been proven to occur in the dentate gyrus, olfactory bulb, and striatum through the measurement of Carbon-14—the levels of which changed during nuclear bomb testing throughout the 20th century—in postmortem human brains. Today, many laboratories around the world are continuing to study the importance of adult neurogenesis in brain function. In 2011, Dr. Altman was awarded the Prince of Asturias Award, an annual prize given in Spain by the Prince of Asturias Foundation to individuals, entities, or organizations globally who make notable achievements in the sciences, humanities, and public affairs. In 2012, he received the International Prize for Biology - an annual award from the Japan Society for the Promotion of Science (JSPS) for "outstanding contribution to the advancement of research in fundamental biology." This Prize is one of the most prestigious honors a scientist can receive. When Dr. Altman died in 2016, Dr. Bayer continued the work they started over 50 years ago. In her late husband's honor, she created the Altman Prize, awarded each year by JSPS to an outstanding young researcher in developmental neuroscience.

INTRODUCTION

A. Specimens and Organization

Volume 13 in the Atlas Series presents the human brain in three normal specimens from the Yakovlev Collection[1] at 9 months at the end of the third trimester. These specimens were analyzed in Volume 2 of the original *Atlas of Human Central Nervous System Development* (Bayer and Altman, 2004). These fetuses are viable *ex utero*. Nearly all the structures present in the adult brain are recognizable and are maturing from the diencephalon to the medulla. But dwindling remnants of the embryonic nervous system remain in the cerebral cortex and the cerebellar cortex.

This volume contains serial grayscale photographs of Nissl-stained sections of a sagittal specimen (Y180-61, **Part II**), a horizontal specimen (Y301-62,

1. The *Yakovlev Collection* (designated by a **Y** prefix in the specimen number) is the work of Dr. Paul Ivan Yakovlev (1894–1983), a neurologist affiliated with Harvard University. Throughout his career, Yakovlev collected many diseased and normal human brains. He invented a giant microtome that was capable of sectioning entire human brains. Later, he became interested in the developing brain and collected many during the second and third trimesters. The normal brains in the developmental group were cataloged by Haleem (1990) and were examined by us during 1996 and 1997. The collection was moved to the National Museum of Health and Medicine when the Armed Forces Institute of Pathology (AFIP) closed at Walter Reed Hospital and is still available for research.

Part III) and a frontal specimen (Y217-65, **Part IV**) with crown-rump (CR) lengths of 310-mm to 350-mm. All specimens are in the 37th gestational week (GW). **Sagittal plates** are ordered from medial to lateral; the anterior part of each photographed section is facing left, posterior right. **Horizontal plates** are presented from dorsal (first) to ventral (last); the anterior part of each section faces left, the posterior part faces right; the midline is in the horizontal center. **Frontal plates** are presented from anterior (first) to posterior (last); the dorsal part of each section is at the top, the ventral part at the bottom; the midline is in the vertical center. Each **Plate** is in two parts: **A**, on the left, shows the full-contrast photograph without labels; **B**, on the right, shows a low-contrast copy of each photograph with unabbreviated labels. For each specimen, a series of serially spaced **low-magnification plates** show the entire section to identify large structures. The brain core is shown in many **high-magnification plates** to identify smaller structures. In addition, several **very-high-magnification plates** show the cerebral and cerebellar cortices. Because our emphasis is on development, transient structures that appear only in immature brains are labeled in *italics*, either directly in some of the high-magnification plates or in **bold numbers** that refer to labels in a list. During dissection, embedding, cutting, and staining, some of the sections illustrated were torn; that damage is sometimes surrounded by *dashed lines*.

B. Developmental Highlights

Figures 1-2 compare early, middle, and late third trimester brains in horizontal sections of the dorsal telencephalon. The overall growth of the brain is obvious when the sections are stacked in a "layer cake" arrangement (**Fig. 1**). The side-by-side comparison (**Fig. 2**) shows that by the end of the third trimester, the cortical plate slightly decreases its thickness and rapidly grows lengthwise. The white matter (*pale yellow*) greatly increases its thickness in most cortical areas except the insula. On the other hand, layers 3-6 of the *stratified transitional field* slow their growth and eventually will disappear. Cortical gyrification continues to elaborate as more fibers enter, and the white matter increases its relative volume. Fibers come from the thalamus (via the enlarging internal capsule) and from the opposite cortex (via the expanding corpus callosum). By now, many associational fibers are accumulating in ipsilateral sides of the cortex to establish connections with neurons in different cortical areas (Altman and Bayer, 2015). Cortical circuitry is continuing to elaborate and the early gyri are more difficult to identify as many secondary and tertiary gyri appear.

Figure 3 provides some insight concerning secondary and tertiary gyrification. The sequence of postnatal myelination gives us a clue to the reason the

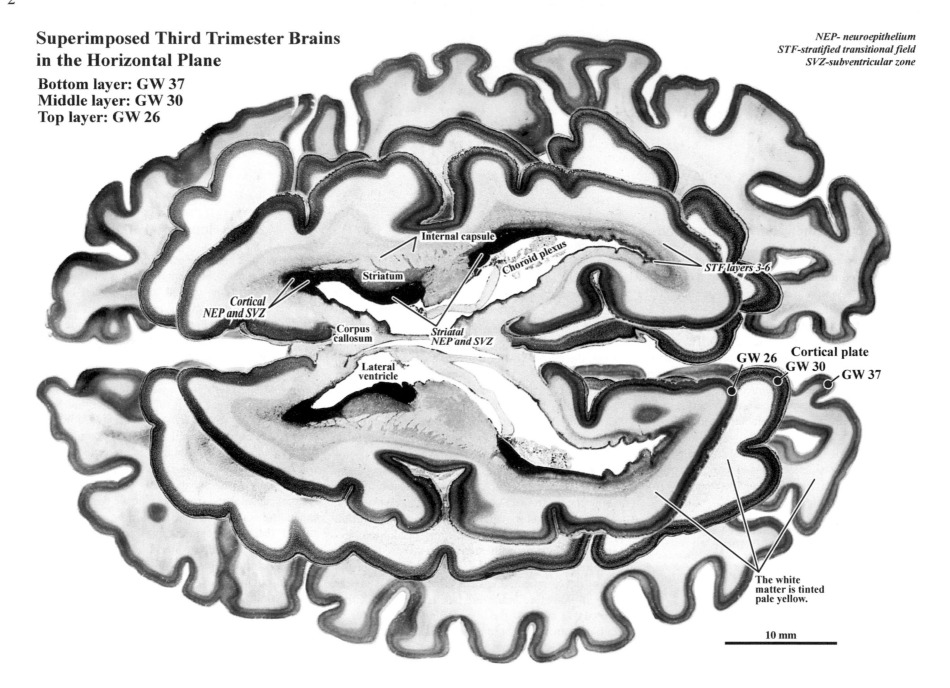

Superimposed Third Trimester Brains in the Horizontal Plane

Bottom layer: GW 37
Middle layer: GW 30
Top layer: GW 26

NEP- neuroepithelium
STF-stratified transitional field
SVZ-subventricular zone

Internal capsule

Striatum

Choroid plexus

STF layers 3-6

Cortical NEP and SVZ

Corpus callosum

Striatal NEP and SVZ

Lateral ventricle

Cortical plate

GW 26
GW 30
GW 37

The white matter is tinted pale yellow.

10 mm

Side-by-Side Comparison of Third Trimester Brains in the Horizontal Plane

GW 37

Growth related to increased gyrification:
Cortical plate *slightly decreases* thickness and *greatly increases* linear surface
White matter increases depth in most areas (insula excepted)
STF layers 3-6 are less prominent and decrease in relative size

GW 30

GW 26

10 mm

Figures 1 and 2 (*facing pages*). A comparison early (GW 26), middle (GW 30), and late (GW 37) trimester brains cut the horizontal plane. All brains are shown at the same scale to emphasize the relative growth of different components of the cerebral cortex. Note that the internal capsule and corpus callosum continue to thicken because many new axons enter. Massive numbers of axons invade the white matter (*pale yellow*). These axons pile up in the white matter and their terminal branches grow into the cortical plate, causing it to thicken and lengthen as synaptic contacts are beginning to be established. That growth results in increased gyrification to the point that the early sulci and gyri are difficult to distinguish. (GW 26 is shown in Plate 28, Volume 11, Bayer and Altman, in press; GW 30 in Plate 58, Volume 12, Bayer and Altman, in press; GW 37 in Plate 21, this volume.)

PROGRESSIVE MYELINATION IN THE CEREBRAL CORTEX
(AFTER THE VOGTS)

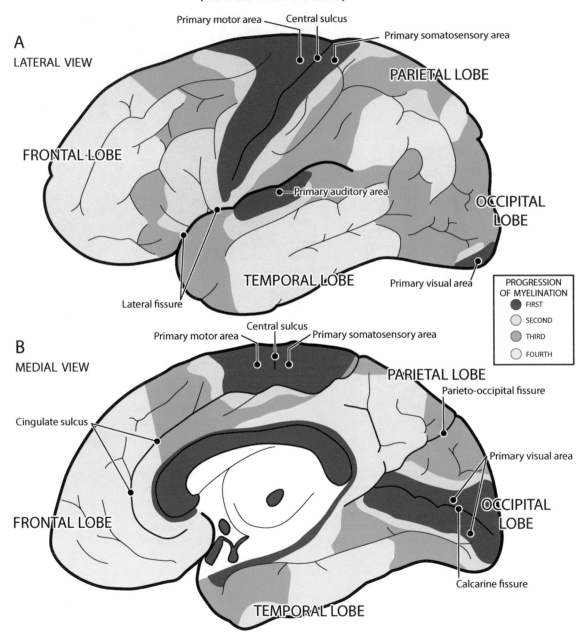

A
LATERAL VIEW

Primary motor area   Central sulcus   Primary somatosensory area

PARIETAL LOBE

FRONTAL LOBE

OCCIPITAL LOBE

Primary auditory area

TEMPORAL LOBE

Primary visual area

Lateral fissure

B
MEDIAL VIEW

Primary motor area   Central sulcus   Primary somatosensory area

PARIETAL LOBE

Parieto-occipital fissure

Cingulate sulcus

Primary visual area

FRONTAL LOBE

OCCIPITAL LOBE

Calcarine fissure

TEMPORAL LOBE

PROGRESSION OF MYELINATION
- FIRST
- SECOND
- THIRD
- FOURTH

primary gyri are "hidden" by secondary and tertiary gyrification. The early gyri and sulci are the primary cortical areas (*red*); note they are a relatively small part of the cortex. Primary areas show little individual variability and are highly conserved in cortical development. They begin to myelinate early in infancy. Secondary areas (*yellow*) surrounding the primary areas myelinate slightly later, adjacent to the primary gyri. Processing here is closely related to processes in the primary areas. Third-order (*dark blue*) and fourth-order (*light blue*) areas mature much later and myelinate last. The late maturing parts of the cortex are association areas where high-order integrative and cognitive processes occur. These areas produce new gyri that mask the prominent appearance of the early gyri in the early third trimester. One can already see the proliferation of new gyri and sulci in the brains presented in this volume, that are near birth. Note that these areas take up the most space in the cortex, and it is here that there is an increase in individual variability. Some parts of the association areas in the frontal lobe (the prefrontal cortex) do not myelinate until the early adult period.

Figure 3. The sequence of myelination based on extensive studies by Vogt and Vogt (1902, 1904). **A**, the lateral cortex, **B**, the medial cortex. Colored areas indicate the myelination sequence: first (*red*), second (*yellow*), third (*dark blue*), fourth (*light blue*). Modified Figure 51 from Altman and Bayer (2015). The original composite figure was first published by Ariéns Kappers et al. (1936).

REFERENCES

Altman J, Bayer SA. (2015) *Development of the Human Neocortex*. Ocala, FL, Laboratory of Developmental Neurobiology, neurondevelopment.org.

Ariéns Kappers CU, Huber GC, Crosby EC (1936) *The Comparative Anatomy of the Nervous System of Vertebrates, Including Man*. 2 volumes. New York, Macmillan (Reprinted in 3 volumes, New York, Hafner, 1960)

Bayer SA, Altman J (2004) *Atlas of Human Central Nervous System Development*, Volume 2: *The Human Bran during the Third Trimester*. Boca Raton, FL, CRC Press.

Bayer SA, Altman J (in press) *The Human Brain during the Second Trimester 225- to 235-mm Crown-Rump Lengths, Atlas of Human Central Nervous System Development*, Volume 11. Taylor and Francis, CRC Press.

Bayer SA, Altman J (in press) *The Human Brain during the Second Trimester 260- to 270-mm Crown-Rump Lengths, Atlas of Human Central Nervous System Development*, Volume 12. Taylor and Francis, CRC Press.

Curtis BA, Jacobson S, Marcus EM (1972) *An Introduction to the Neurosciences*, Philadelphia: W. B. Saunders.

Haleem M (1990) *Diagnostic Categories of the Yakovlev Collection of Normal and Pathological Anatomy and Development of the Brain*. Washington, D.C. Armed Forces Institute of Pathology.

Larroche JC (1966) The development of the central nervous system during intrauterine life. In: *Human Development*, F. Falkner (ed.), Philadelphia: W. B. Saunders, pages 257-276.

Vogt C, Vogt O (1902) *Bieträge zur Hirnfasernlehre* Volume 1, *Zur Erforschung der Hirnfaserung*. Jena, Fischer.

Vogt C, Vogt O (1904) *Die Markretfung des Kindergehirns während der ersten vier Lebensmonate und thre metholologische Bedeutung*, Volume 2, Jena, Fischer.

PART II: Y180-61
CR 310 mm (GW 37)
Sagittal

This specimen is case number B-180-61 (Perinatal RPSL) in the Yakovlev Collection. A female infant survived for one hour after birth. Death occurred because a hyaline membrane obstructed the airway to the lungs. The brain was cut in the sagittal plane in 35-μm thick sections and is classified as a Normative Control in the Yakovlev Collection (Haleem, 1990). There is no photograph of this brain before it was embedded and cut, so the photograph of another GW 37 brain that Larroche published in 1966 (**Figure 4**) is used to show medial surface features.

Photographs of 8 Nissl-stained sections are shown in **Plates 1-8**. The core of the brain and the cerebellum are shown at higher magnification in **Plates 9–16**. Very-high-magnification views of different regions of the cerebellar cortex are shown in **Plates 17** and **18**. Because the section numbers decrease from **Plate 1** (most medial) to **Plate 8** (most lateral), they are from the left side of the brain; the right side has higher section numbers proceeding medial to lateral. The cutting plane of this brain is nearly parallel to the midline. However, the posterior part of each section is angled more toward the right than the anterior part. For example, in **Plate 1** the paraflocculus from the right side of the brain is beneath the cerebellar vermis. The paraflocculus is very small in **Plate 2** and the left

paraflocculus appears in **Plate 3**. The sections chosen for illustration are spaced closer together near the midline to show small structures in the diencephalon, midbrain, pons, and medulla.

Y180-61 contains remnants of the germinal matrices in all lobes of the cerebral cortex where the *neuroepithelium/subventricular zone* is still generating neocortical interneurons. Migrating and sojourning neurons and/or glia are in the dwindling *stratified transitional fields* in all lobes of the cerebral cortex, more prominent in sensory rather than motor cortical areas.

Many neurons, glia, and their mitotic precursor cells are still migrating through the olfactory peduncle toward the olfactory bulb in the *rostral migratory stream* from a presumed source area in the germinal matrix at the junction between the cerebral cortex, striatum, and nucleus accumbens. Lateral parts of the cerebral cortex has streams of neurons and glia in the *lateral migratory stream* that percolates through the claustrum, endopiriform nucleus, external capsule, and uncinate fasciculus. These cells appear to be heading toward the insular cortex, primary olfactory cortex, temporal cortex, and basolateral parts of the amygdaloid complex.

In the basal ganglia, a prominent *neuroepithelium/ subventricular zone* covers the striatum and nucleus accumbens where neurons are being generated. The *subgranular zone* in the hilus of the dentate gyrus is generating granule cells. Other structures in the telencephalon, such as the septum, fornix, and Ammon's horn part of the hippocampus, have only a densely stained layer at the ventricle; presumably generating glia, cells of the choroid plexus, and the ependymal lining of the ventricle.

Most of the structures in the diencephalon appear to be settled and maturing; the third ventricle is lined by a thin *glioepithelium/ependyma*. In the midbrain and anterior pons, there is a slightly thicker and more convoluted *glioepithelium/ependyma* lining the posterior cerebral aqueduct and anterior fourth ventricle. The posterior pons and entire medulla have a thin *glioepithelium/ependyma* lining the rest of the fourth ventricle.

The *external germinal layer* is prominent over the entire surface of the cerebellar cortex and is producing basket, stellate, and granule cells. The *germinal trigone* is still visible at the base of the nodulus and along the floccular peduncle; choroid plexus cells and glia are originating here.

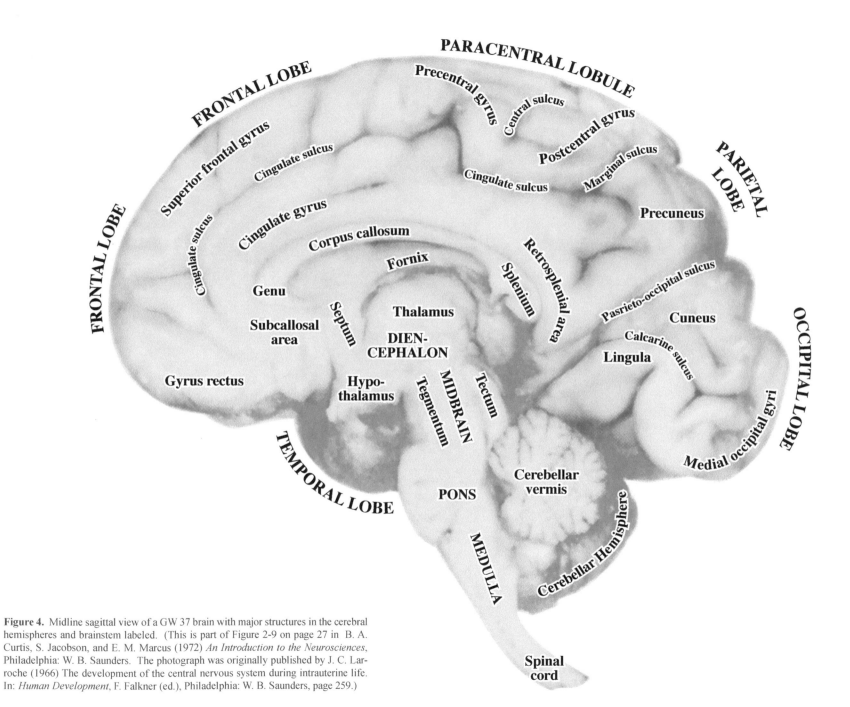

Figure 4. Midline sagittal view of a GW 37 brain with major structures in the cerebral hemispheres and brainstem labeled. (This is part of Figure 2-9 on page 27 in B. A. Curtis, S. Jacobson, and E. M. Marcus (1972) *An Introduction to the Neurosciences*, Philadelphia: W. B. Saunders. The photograph was originally published by J. C. Larroche (1966) The development of the central nervous system during intrauterine life. In: *Human Development*, F. Falkner (ed.), Philadelphia: W. B. Saunders, page 259.)

PLATE 1A
CR 310 mm
GW 37, Y180-61
Sagittal
Section 961

See detail of the brain core
and cerebellum in
Plates 9A and B.

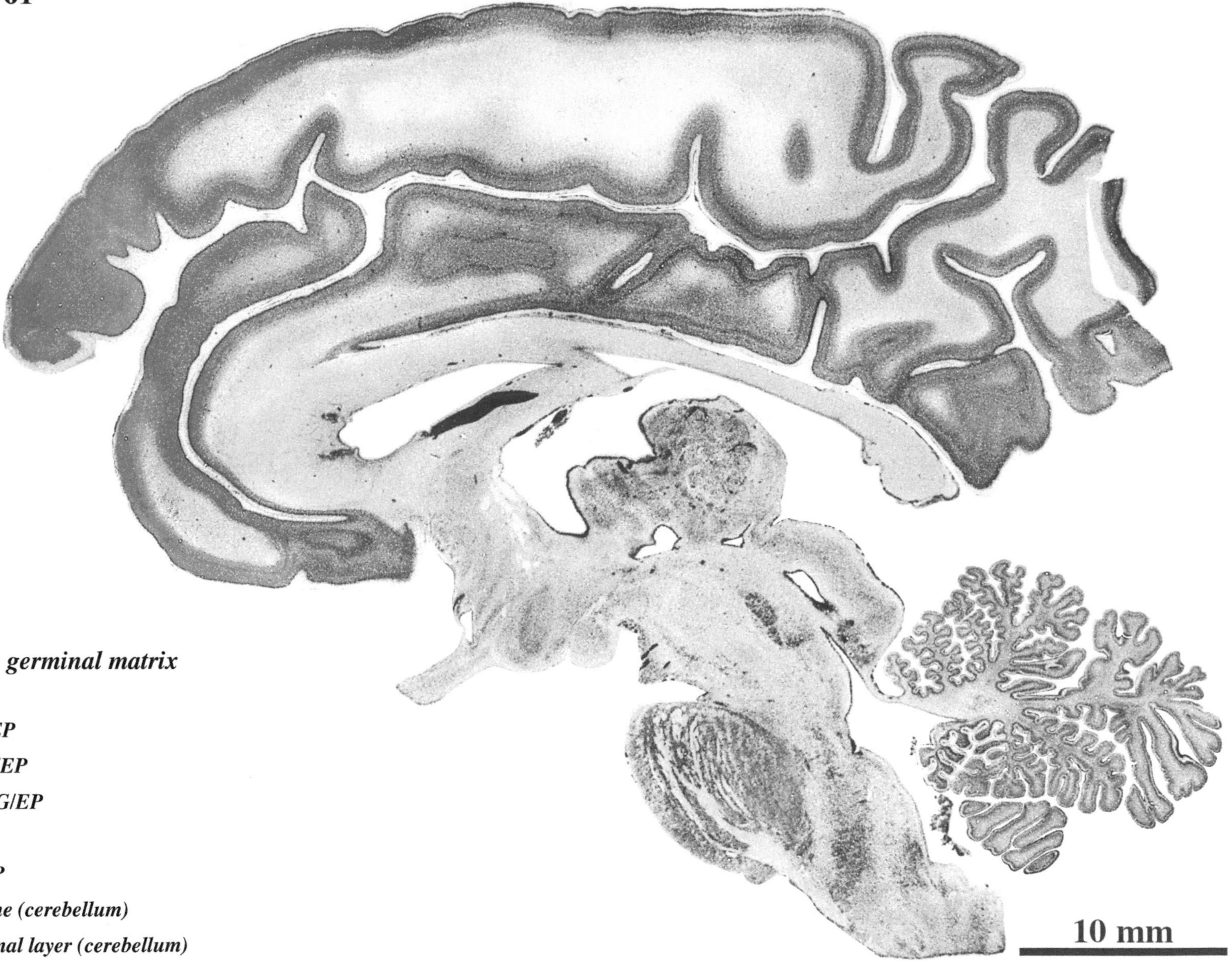

Remnants of the germinal matrix

1 *Callosal sling*

2 *Strionuclear GEP*

3 *Diencephalic G/EP*

4 *Mesencephalic G/EP*

5 *Pontine G/EP*

6 *Medullary G/EP*

7 *Germinal trigone (cerebellum)*

8 *External germinal layer (cerebellum)*

9 *Subpial granular layer (cortical)*

GEP - Glioepithelium
G/EP - Glioepithelium/ependyma

10 mm

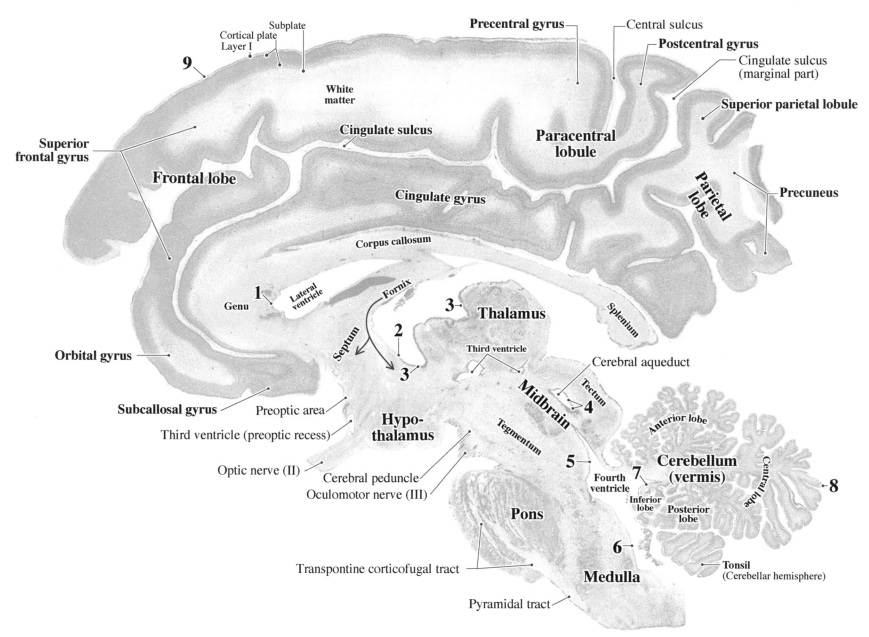

Subplate

Cortical plate
Layer I

9

White
matter

Precentral gyrus

Central sulcus

Postcentral gyrus

Cingulate sulcus
(marginal part)

Superior parietal lobule

**Superior
frontal gyrus**

Cingulate sulcus

Frontal lobe

**Paracentral
lobule**

**Parietal
lobe**

Cingulate gyrus

Precuneus

Corpus callosum

1

Genu

Lateral
ventricle

Fornix

3

Thalamus

Splenium

2

Septum

3

Third ventricle

Cerebral aqueduct

Orbital gyrus

Tectum

Midbrain

4

Subcallosal gyrus

Preoptic area

Third ventricle (preoptic recess)

**Hypo-
thalamus**

Tegmentum

Anterior lobe

Optic nerve (II)

Cerebral peduncle

Oculomotor nerve (III)

5

Fourth
ventricle

7

**Cerebellum
(vermis)**

Central lobe

8

Inferior
lobe

Posterior
lobe

Pons

6

Transpontine corticofugal tract

Medulla

Tonsil
(Cerebellar hemisphere)

Pyramidal tract

PLATE 2A
CR 310 mm
GW 37, Y180-61
Sagittal
Section 941

See detail of the brain core
and cerebellum in
Plates 10A and B.

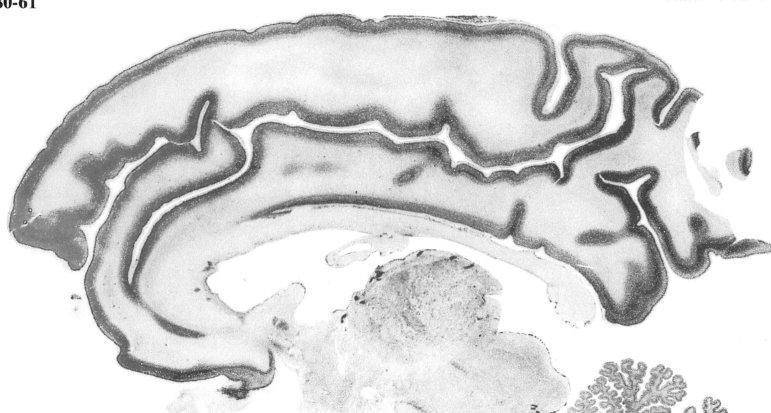

Remnants of the germinal matrix

1 *Callosal sling*

2 *Strionuclear GEP*

3 *Mesencephalic G/EP*

4 *Pontine G/EP*

5 *Medullary G/EP*

6 *Germinal trigone*

7 *External germinal layer (cerebellum)*

8 *Subpial granular layer (cortical)*

GEP - Glioepithelium
G/EP - Glioepithelium/ependyma

10 mm

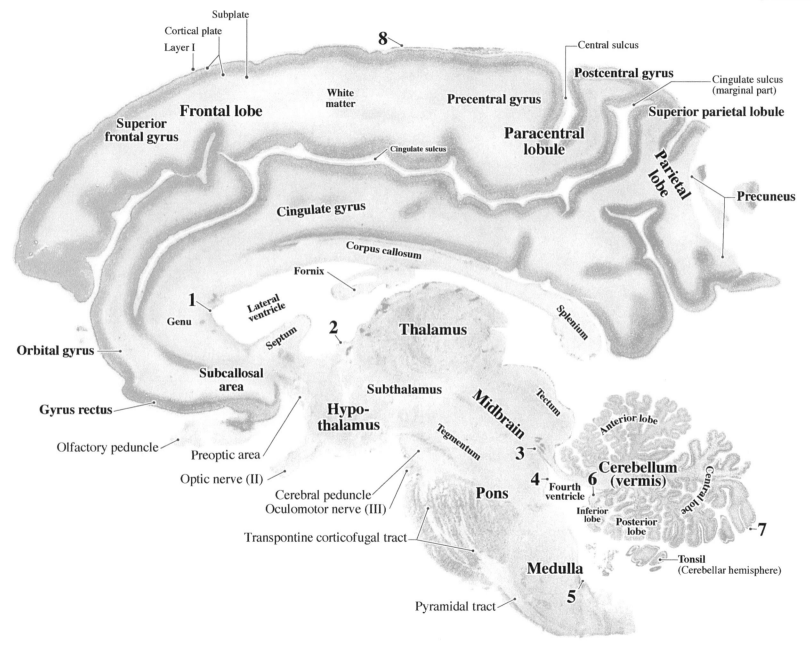

Subplate

Cortical plate

Layer I

8

Central sulcus

Postcentral gyrus

Cingulate sulcus (marginal part)

White matter

Precentral gyrus

Frontal lobe

Paracentral lobule

Superior parietal lobule

Superior frontal gyrus

Cingulate sulcus

Parietal lobe

Precuneus

Cingulate gyrus

Corpus callosum

Fornix

1

Lateral ventricle

Splenium

Genu

2

Thalamus

Septum

Orbital gyrus

Subcallosal area

Subthalamus

Midbrain

Tectum

Anterior lobe

Gyrus rectus

Hypothalamus

Tegmentum

3

Cerebellum (vermis)

Central lobe

Olfactory peduncle

Preoptic area

4 Fourth ventricle

6

Optic nerve (II)

Cerebral peduncle
Oculomotor nerve (III)

Pons

Inferior lobe

Posterior lobe

7

Transpontine corticofugal tract

Medulla

Tonsil (Cerebellar hemisphere)

Pyramidal tract

5

PLATE 3A
CR 310 mm
GW 37, Y180-61
Sagittal
Section 841

See detail of the brain core
and cerebellum in
Plates 11A and B.

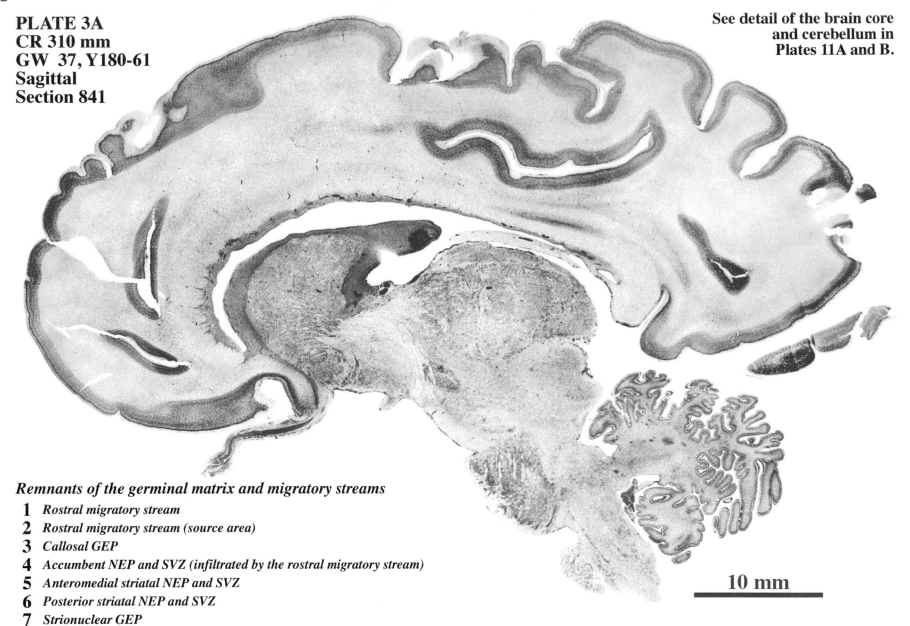

Remnants of the germinal matrix and migratory streams

1 *Rostral migratory stream*
2 *Rostral migratory stream (source area)*
3 *Callosal GEP*
4 *Accumbent NEP and SVZ (infiltrated by the rostral migratory stream)*
5 *Anteromedial striatal NEP and SVZ*
6 *Posterior striatal NEP and SVZ*
7 *Strionuclear GEP*
8 *Pontine and medullary G/EP*
9 *External germinal layer (cerebellum)*
10 *Subpial granular layer (cortical)*

GEP - Glioepithelium
G/EP - Glioepithelium/ependyma
NEP - Neuroepithelium
SVZ - Subventricular zone

10 mm

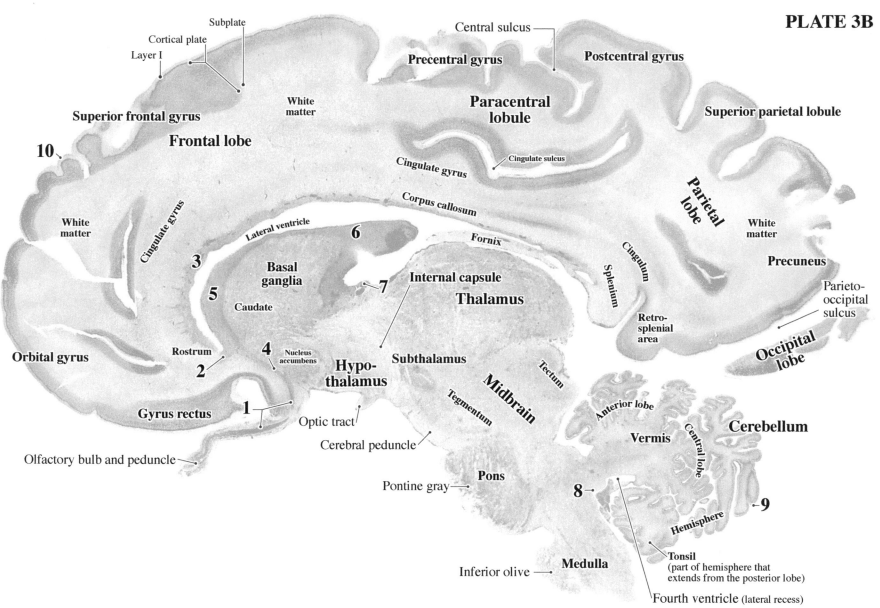

Subplate

Cortical plate

Layer I

Central sulcus

Precentral gyrus

Postcentral gyrus

White matter

Paracentral lobule

Superior parietal lobule

Superior frontal gyrus

Frontal lobe

Cingulate gyrus

Cingulate sulcus

10

Cingulate gyrus

Corpus callosum

Parietal lobe

White matter

Lateral ventricle

6

Fornix

Cingulum

Precuneus

3

Basal ganglia

Splenium

White matter

5

Caudate

7

Internal capsule

Thalamus

Retro-splenial area

Parieto-occipital sulcus

Orbital gyrus

Rostrum

4

Nucleus accumbens

2

Subthalamus

Tectum

Occipital lobe

1

Hypo-thalamus

Midbrain

Anterior lobe

Cerebellum

Gyrus rectus

Optic tract

Tegmentum

Vermis

Central lobe

Cerebral peduncle

Olfactory bulb and peduncle

Hemisphere

9

Pontine gray

Pons

8

Inferior olive

Medulla

Tonsil
(part of hemisphere that extends from the posterior lobe)

Fourth ventricle (lateral recess)

**PLATE 4A
CR 310 mm
GW 37, Y180-61
Sagittal
Section 781**

See detail of the brain core
and cerebellum in
Plates 12A and B.

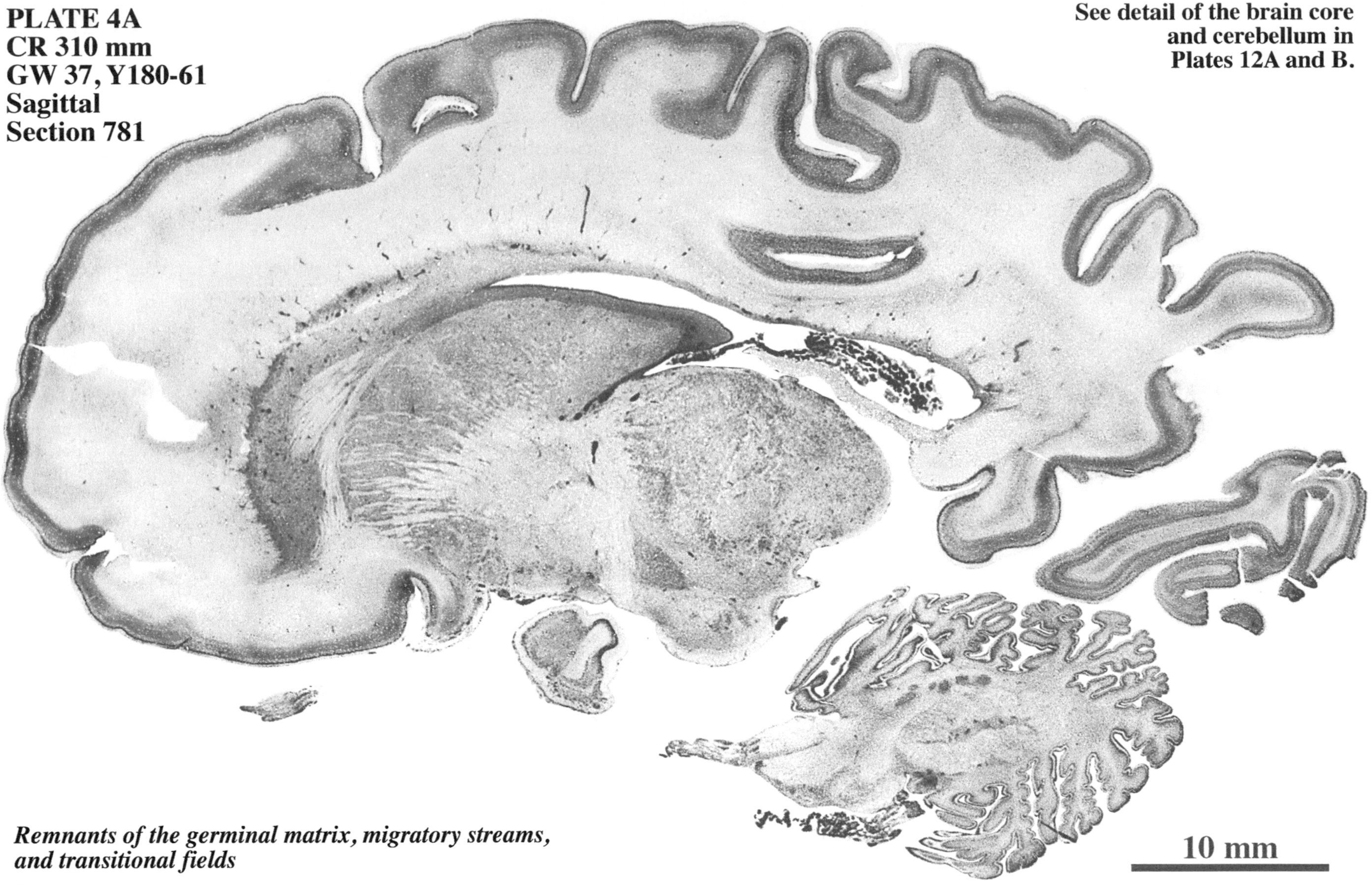

*Remnants of the germinal matrix, migratory streams,
and transitional fields*

1 *Rostral migratory stream*
2 *Frontal NEP and SVZ*
3 *Frontal STF*
4 *Callosal GEP*
5 *Fornical GEP*

6 *Anterolateral striatal NEP and SVZ*
7 *Posterior striatal NEP and SVZ*
8 *Strionuclear GEP*
9 *External germinal layer (cerebellum)*
10 *Subpial granular layer (cortical)*

GEP - Glioepithelium
NEP - Neuroepithelium
STF - Stratified transitional field
SVZ - Subventricular zone

10 mm

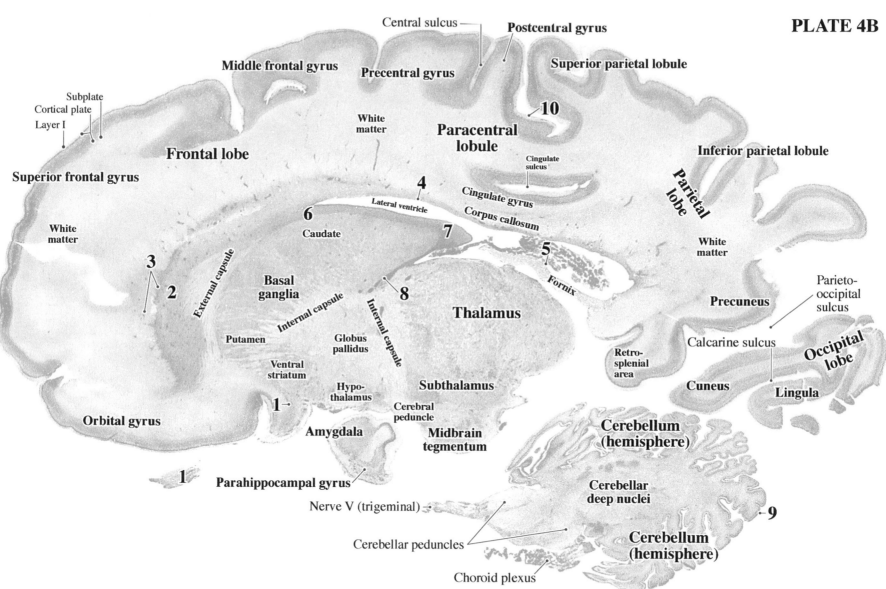

Central sulcus — **Postcentral gyrus**

Middle frontal gyrus **Precentral gyrus** **Superior parietal lobule**

Subplate
Cortical plate
Layer I

White
matter

**Paracentral
lobule**

10

Inferior parietal lobule

Frontal lobe

Cingulate
sulcus

4

Cingulate gyrus

**Parietal
lobe**

Superior frontal gyrus

6

Lateral ventricle

Corpus callosum

White
matter

Caudate

7

5

White
matter

3

2

External capsule

**Basal
ganglia**

Internal capsule

8

Fornix

Precuneus

Parieto-
occipital
sulcus

Internal capsule

Thalamus

Calcarine sulcus

**Occipital
lobe**

Putamen

Globus
pallidus

Retro-
splenial
area

Cuneus

Lingula

Ventral
striatum

Hypo-
thalamus

Subthalamus

1

Cerebral
peduncle

Orbital gyrus

Amygdala

Cerebral
peduncle

**Midbrain
tegmentum**

**Cerebellum
(hemisphere)**

1

Parahippocampal gyrus

**Cerebellar
deep nuclei**

9

Nerve V (trigeminal)

**Cerebellum
(hemisphere)**

Cerebellar peduncles

Choroid plexus

PLATE 5A
CR 310 mm
GW 37, Y180-61
Sagittal
Section 721

See detail of the brain core
and cerebellum in
Plates 13A and B.

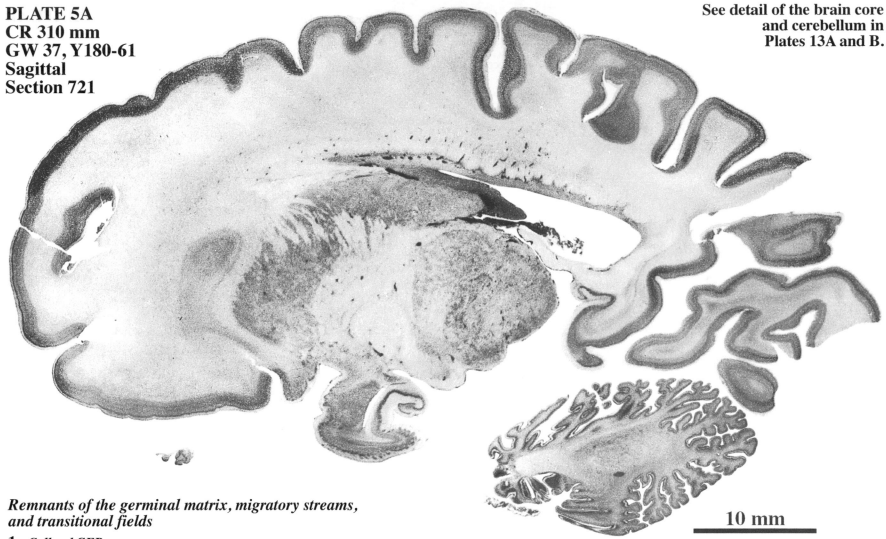

10 mm

Remnants of the germinal matrix, migratory streams,
and transitional fields

1 *Callosal GEP*

2 *Fornical GEP*

3 *Alvear GEP*

4 *Subgranular zone (dentate)*

5 *Lateral migratory stream (cortical)*

6 *Amygdaloid G/EP*

7 *Anterolateral striatal NEP and SVZ*

8 *Posterior striatal NEP and SVZ*

9 *Strionuclear GEP*

10 *External germinal layer (cerebellum)*

11 *Subpial granular layer (cortical)*

GEP - Glioepithelium
G/EP - Glioepithelium/ependyma
NEP - Neuroepithelium
SVZ - Subventricular zone

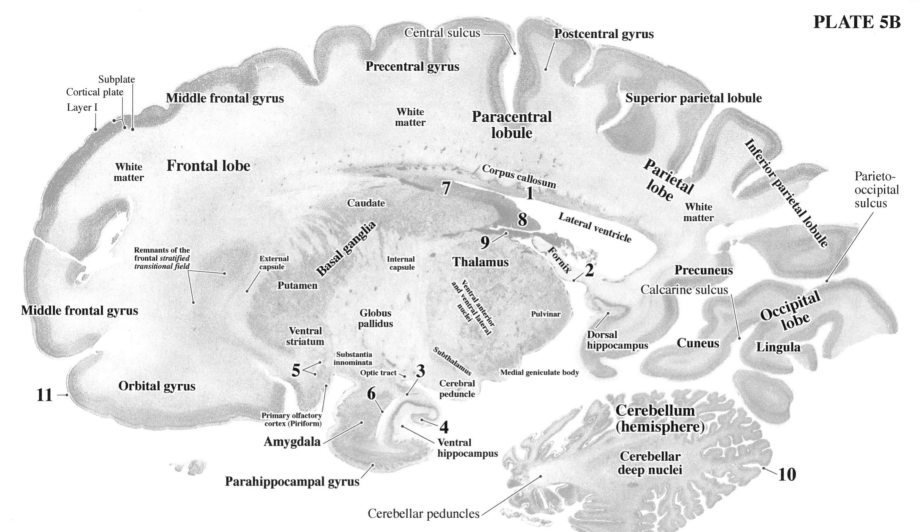

Central sulcus

Postcentral gyrus

Precentral gyrus

Subplate

Cortical plate

Layer I

Middle frontal gyrus

White matter

Paracentral lobule

Superior parietal lobule

White matter

Frontal lobe

Corpus callosum

Parietal lobe

White matter

Inferior parietal lobule

Parieto-occipital sulcus

7

1

Caudate

8

Lateral ventricle

9

2

Remnants of the frontal *stratified transitional field*

External capsule

Basal ganglia

Internal capsule

Thalamus

Fornix

Precuneus

Putamen

Ventral anterior and ventral lateral nuclei

Pulvinar

Calcarine sulcus

Occipital lobe

Middle frontal gyrus

Globus pallidus

Dorsal hippocampus

Cuneus

Lingula

Ventral striatum

Substantia innominata

Subthalamus

Medial geniculate body

5

Optic tract

3

Cerebral peduncle

6

11

Orbital gyrus

Primary olfactory cortex (Piriform)

4

Ventral hippocampus

Cerebellum (hemisphere)

Cerebellar deep nuclei

10

Amygdala

Parahippocampal gyrus

Cerebellar peduncles

PLATE 6A
CR 310 mm
GW 37, Y180-61
Sagittal
Section 661

See detail of the brain core
and cerebellum in
Plates 14A and B.

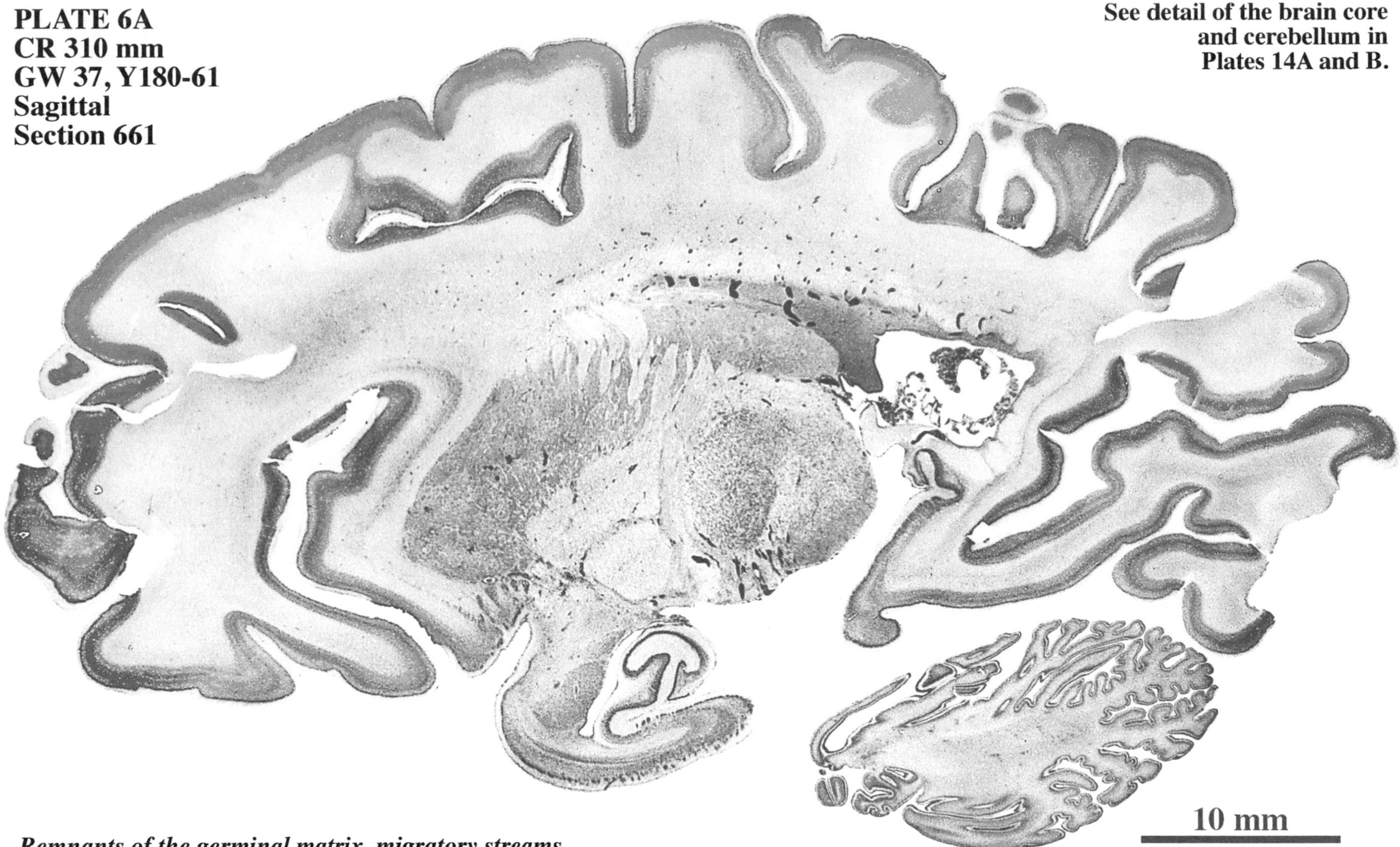

10 mm

Remnants of the germinal matrix, migratory streams,
and transitional fields

1 *Callosal GEP*
2 *Fornical GEP*
3 *Alvear GEP*
4 *Subgranular zone (dentate)*
5 *Lateral migratory stream (cortical)*

6 *Amygdaloid G/EP*
7 *Posterior striatal NEP and SVZ*
8 *Strionuclear GEP*
9 *External germinal layer (cerebellum)*
10 *Subpial granular layer (cortical)*

GEP - Glioepithelium
G/EP - Glioepithelium/ependyma
NEP - Neuroepithelium
SVZ - Subventricular zone

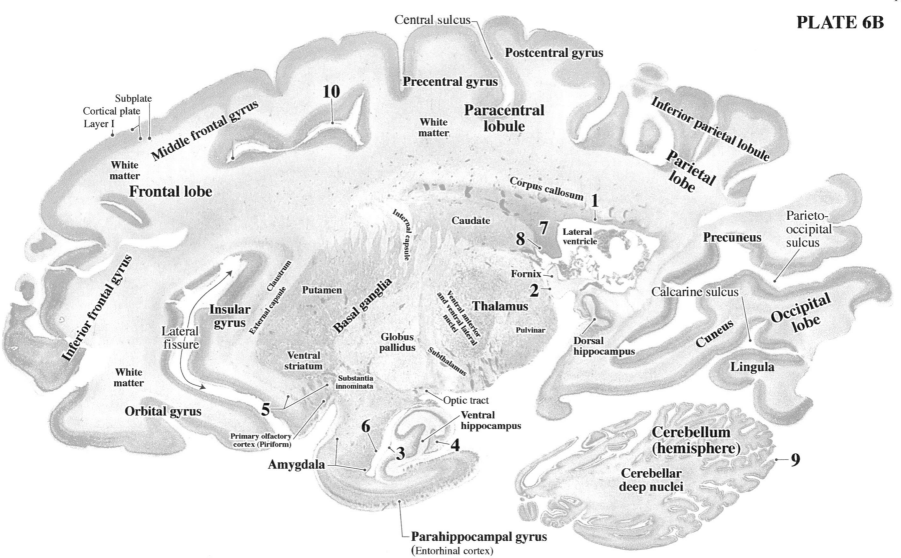

Central sulcus

Postcentral gyrus

Precentral gyrus

10

Paracentral lobule

White matter

Inferior parietal lobule

Subplate

Cortical plate

Layer I

Middle frontal gyrus

White matter

Frontal lobe

Parietal lobe

Corpus callosum

1

Caudate

7

8

Lateral ventricle

Precuneus

Parieto-occipital sulcus

Internal capsule

Fornix

2

Inferior frontal gyrus

Clastrum

Putamen

Basal ganglia

Ventral anterior and ventral lateral nuclei

Thalamus

Calcarine sulcus

Occipital lobe

Insular gyrus

External capsule

Pulvinar

Dorsal hippocampus

Cuneus

Lateral fissure

Globus pallidus

Subthalamus

Lingula

White matter

Ventral striatum

Substantia innominata

Optic tract

Ventral hippocampus

Orbital gyrus

5

6

3

4

Cerebellum (hemisphere)

9

Primary olfactory cortex (Piriform)

Amygdala

Cerebellar deep nuclei

Parahippocampal gyrus
(Entorhinal cortex)

PLATE 7A
CR 310 mm
GW 37, Y180-61
Sagittal
Section 621

See detail of the brain core
and cerebellum in
Plates 15A and B.

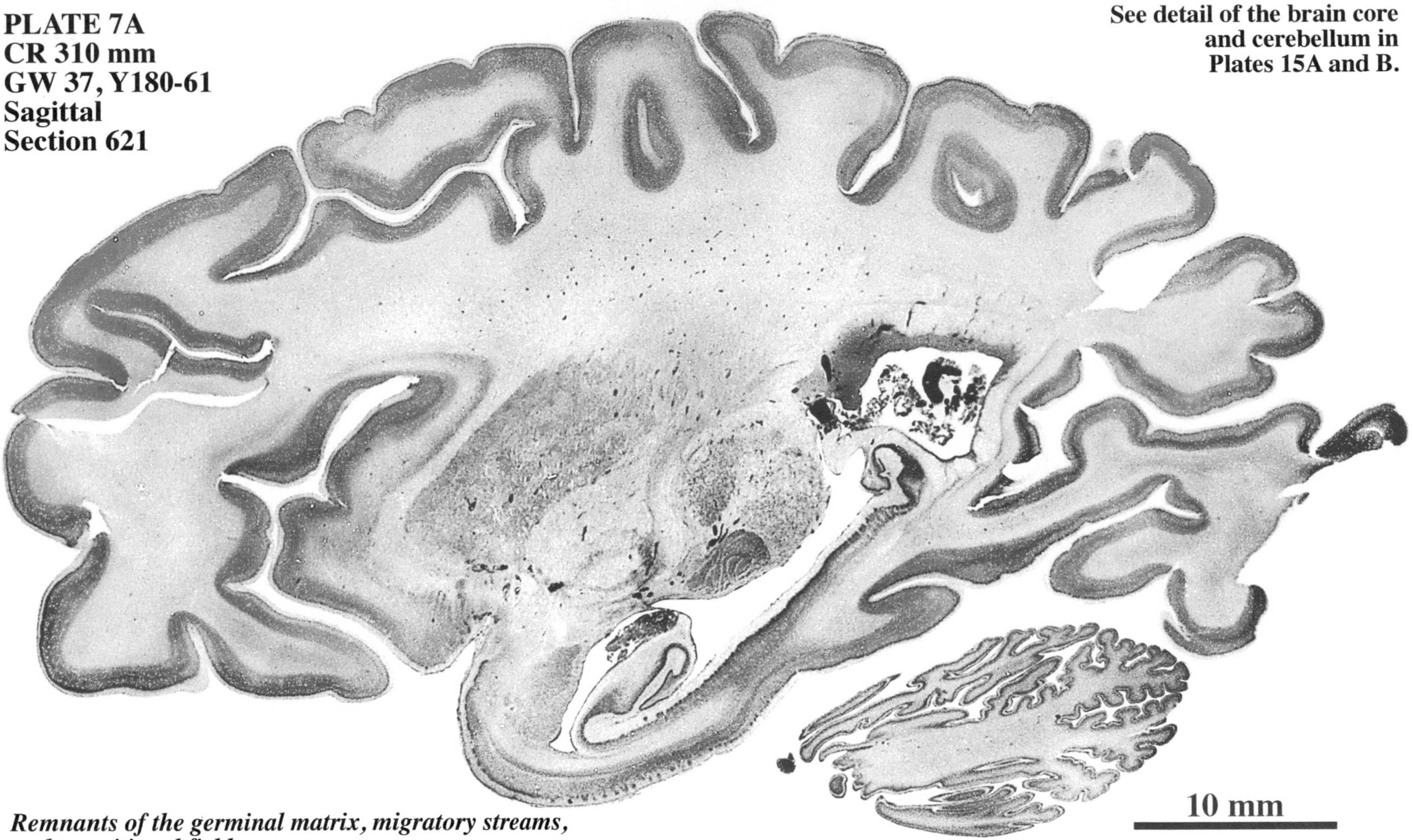

10 mm

Remnants of the germinal matrix, migratory streams,
and transitional fields

1 *Callosal GEP*

2 *Fornical GEP*

3 *Parahippocampal NEP, SVZ, and STF*

4 *Alvear GEP*

5 *Subgranular zone (dentate)*

6 *Lateral migratory stream (cortical)*

7 *Amygdaloid G/EP*

8 *Posterior striatal NEP and SVZ*

9 *Strionuclear GEP*

10 *External germinal layer (cerebellum)*

11 *Subpial granular layer (cortical)*

GEP - Glioepithelium
G/EP - Glioepithelium/ependyma
NEP - Neuroepithelium
STF - Stratified transitional field
SVZ - Subventricular zone

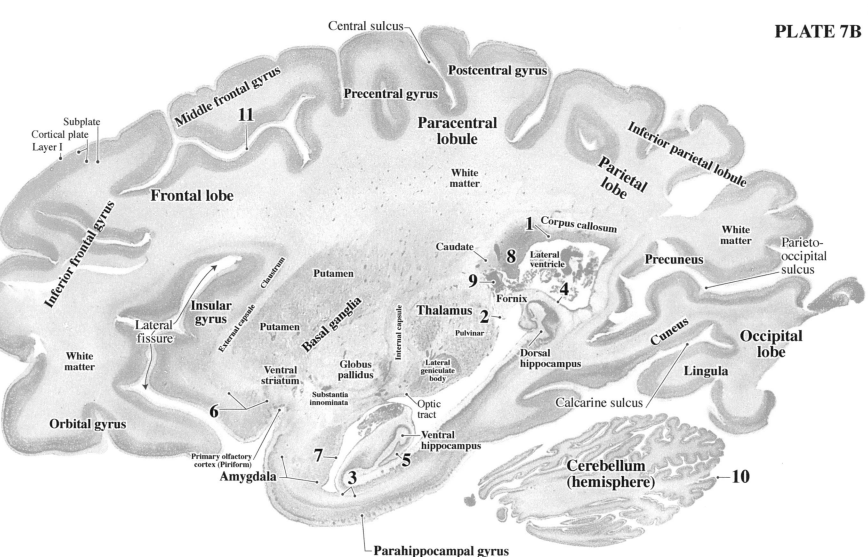

Central sulcus

Postcentral gyrus

Precentral gyrus

Middle frontal gyrus **11**

Paracentral lobule

Parietal lobe

Inferior parietal lobule

Subplate
Cortical plate
Layer I

White matter

Frontal lobe

Inferior frontal gyrus

1 Corpus callosum

Caudate

8 Lateral ventricle

Precuneus

White matter

Parieto-occipital sulcus

9

4

Putamen

Claustrum

Fornix

Insular gyrus

External capsule

Lateral fissure

Putamen

Basal ganglia

Internal capsule

Thalamus **2**

Pulvinar

Dorsal hippocampus

Cuneus

Occipital lobe

White matter

Globus pallidus

Lateral geniculate body

Lingula

Ventral striatum

Substantia innominata

Optic tract

Calcarine sulcus

Orbital gyrus

Ventral hippocampus

6

Primary olfactory cortex (Piriform)

7

5

Cerebellum (hemisphere)

10

Amygdala

3

Parahippocampal gyrus
(Entorhinal cortex)

PLATE 8A
CR 310 mm
GW 37, Y180-61
Sagittal
Section 501

See detail of the brain core
and cerebellum in
Plates 16A and B.

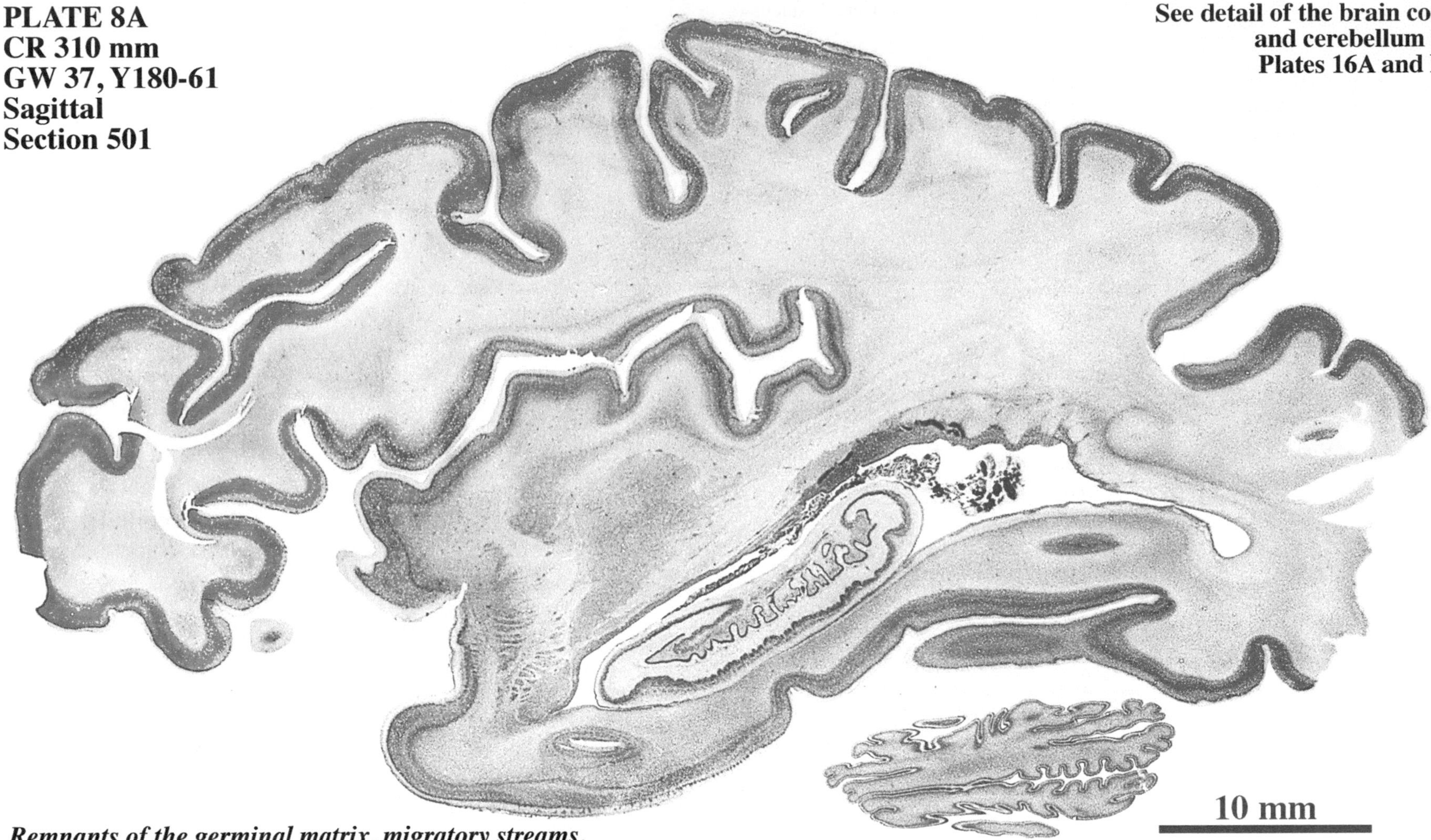

10 mm

Remnants of the germinal matrix, migratory streams,
and transitional fields

1 *Parietal NEP, SVZ, and STF*

2 *Occipital NEP, SVZ, and STF*

3 *Parahippocampal NEP, SVZ, and STF*

4 *Alvear GEP*

5 *Subgranular zone (dentate)*

6 *Lateral migratory stream (cortical)*

7 *Amygdaloid G/EP*

8 *Posterior striatal NEP and SVZ*

9 *Strionuclear GEP*

10 *External germinal layer (cerebellum)*

11 *Subpial granular layer (cortical)*

GEP - Glioepithelium
G/EP - Glioepithelium/ependyma
NEP - Neuroepithelium
STF - Stratified transitional field
SVZ - Subventricular zone

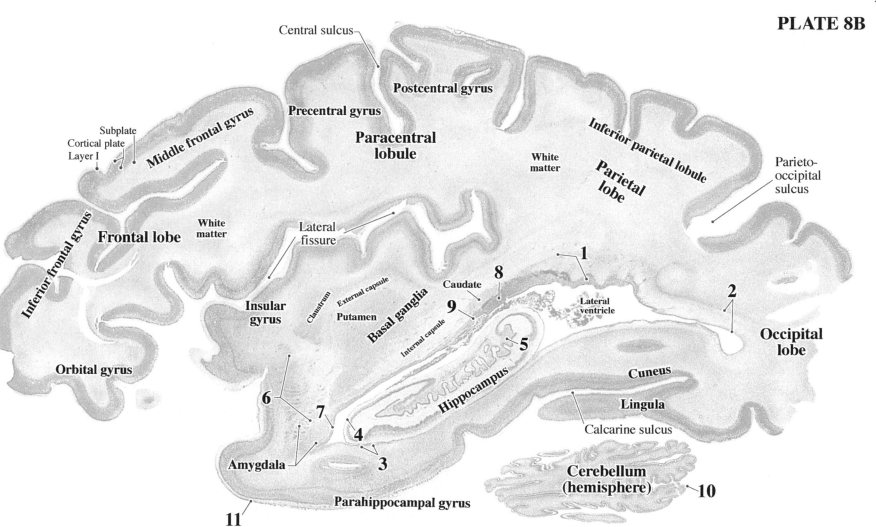

Central sulcus

Postcentral gyrus

Precentral gyrus

Paracentral lobule

Subplate
Cortical plate
Layer I

Middle frontal gyrus

Inferior parietal lobule

White matter

Parietal lobe

Parieto-occipital sulcus

Inferior frontal gyrus

Frontal lobe

White matter

Lateral fissure

1

8

Caudate

Lateral ventricle

2

Claustrum
External capsule

Insular gyrus

Putamen

Basal ganglia

9

Occipital lobe

Internal capsule

5

Orbital gyrus

6

7

Hippocampus

Cuneus

Lingula

Calcarine sulcus

4

3

Amygdala

Cerebellum (hemisphere)

10

11

Parahippocampal gyrus

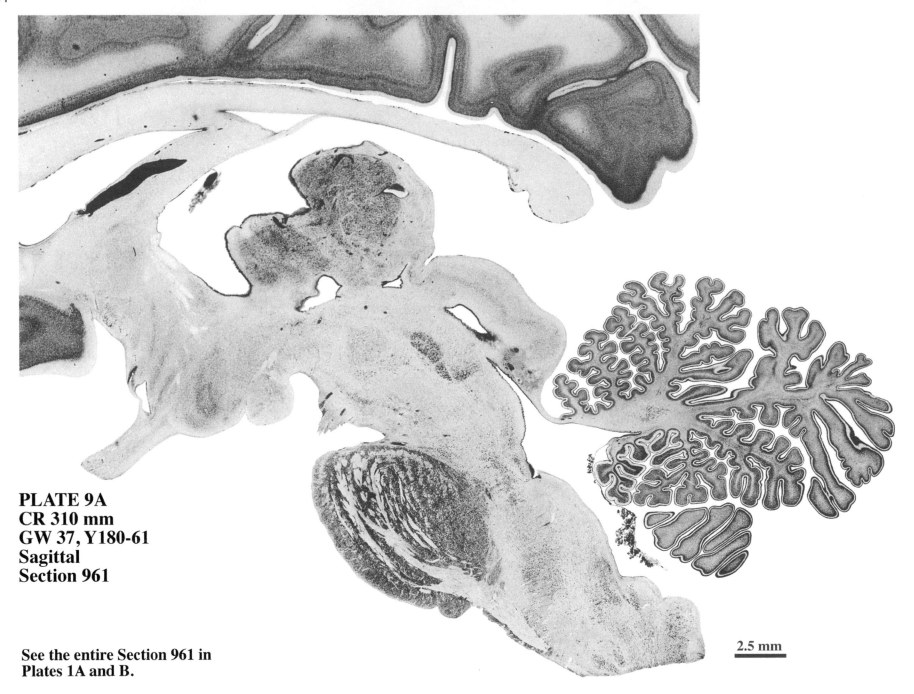

PLATE 9A
CR 310 mm
GW 37, Y180-61
Sagittal
Section 961

See the entire Section 961 in
Plates 1A and B.

2.5 mm

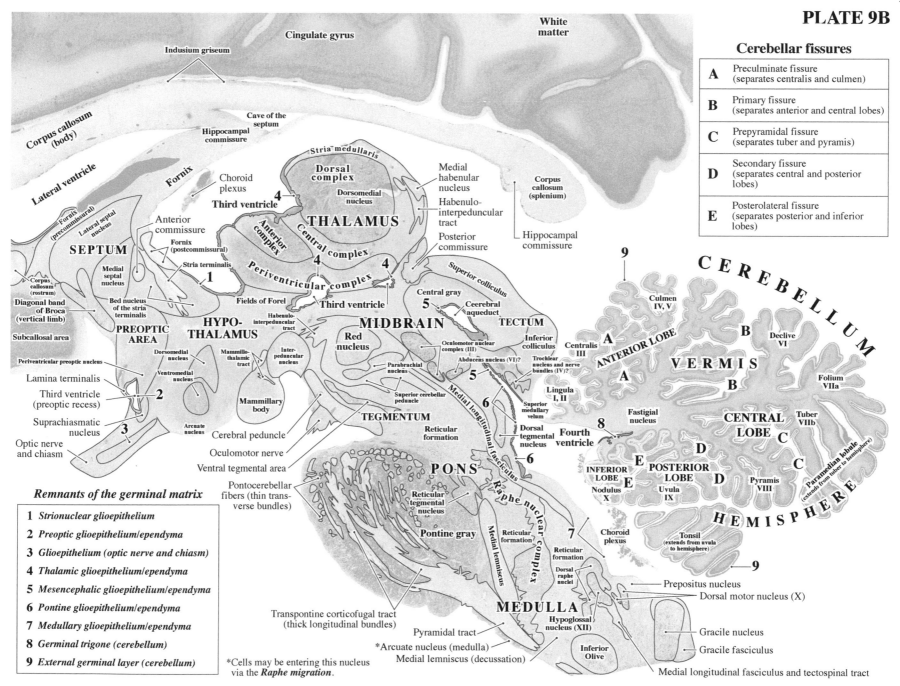

PLATE 9B

Cerebellar fissures

A	Preculminate fissure (separates centralis and culmen)
B	Primary fissure (separates anterior and central lobes)
C	Prepyramidal fissure (separates tuber and pyramis)
D	Secondary fissure (separates central and posterior lobes)
E	Posterolateral fissure (separates posterior and inferior lobes)

Cingulate gyrus

Indusium griseum

White matter

Corpus callosum (body)

Cave of the septum

Hippocampal commissure

Stria medullaris

Dorsal complex

Medial habenular nucleus

Habenulo-interpeduncular tract

Corpus callosum (splenium)

Lateral ventricle

Fornix

Choroid plexus

Third ventricle

Dorsomedial nucleus

THALAMUS

Posterior commissure

Hippocampal commissure

Fornix (precommissural)

Lateral septal nucleus

Anterior commissure

Anterior complex

Central complex

Superior colliculus

SEPTUM

Medial septal nucleus

Fornix (postcommissural)

Stria terminalis

Periventricular complex

Central gray

Cerebral aqueduct

TECTUM

Corpus callosum (rostrum)

Bed nucleus of the stria terminalis

Fields of Forel

Third ventricle

MIDBRAIN

Inferior colliculus

Centralis III

ANTERIOR LOBE

Culmen IV, V

Diagonal band of Broca (vertical limb)

HYPO-THALAMUS

Habenulo-interpeduncular tract

Red nucleus

Oculomotor nuclear complex (III)

Abducens nucleus (VI)?

Trochlear nucleus and nerve bundles (IV)?

VERMIS

Declive VI

Subcallosal area

PREOPTIC AREA

Dorsomedial nucleus

Mammillo-thalamic tract

Inter-peduncular nucleus

Parabrachial nucleus

Superior cerebellar peduncle

Lingula I, II

Folium VIIa

Periventricular preoptic nucleus

Ventromedial nucleus

Superior medullary velum

Fastigial nucleus

CENTRAL LOBE

Tuber VIIb

Lamina terminalis

Mammillary body

TEGMENTUM

Reticular formation

Dorsal tegmental nucleus

Fourth ventricle

Third ventricle (preoptic recess)

Arcuate nucleus

Cerebral peduncle

INFERIOR LOBE

POSTERIOR LOBE

Pyramis VIII

Suprachiasmatic nucleus

Oculomotor nerve

Ventral tegmental area

PONS

Nodulus X

Uvula IX

Optic nerve and chiasm

Pontocerebellar fibers (thin transverse bundles)

Reticular tegmental nucleus

HEMISPHERE

Tonsil (extends from uvula to hemisphere)

Pontine gray

Reticular formation

Choroid plexus

Remnants of the germinal matrix

1 *Strionuclear glioepithelium*

2 *Preoptic glioepithelium/ependyma*

3 *Glioepithelium (optic nerve and chiasm)*

4 *Thalamic glioepithelium/ependyma*

5 *Mesencephalic glioepithelium/ependyma*

6 *Pontine glioepithelium/ependyma*

7 *Medullary glioepithelium/ependyma*

8 *Germinal trigone (cerebellum)*

9 *External germinal layer (cerebellum)*

Medial lemniscus

Reticular formation

Dorsal raphe nuclei

MEDULLA

Hypoglossal nucleus (XII)

Prepositus nucleus

Dorsal motor nucleus (X)

Transpontine corticofugal tract (thick longitudinal bundles)

Pyramidal tract

*Arcuate nucleus (medulla)

Medial lemniscus (decussation)

Inferior Olive

Gracile nucleus

Gracile fasciculus

Medial longitudinal fasciculus and tectospinal tract

*Cells may be entering this nucleus via the **Raphe migration**.

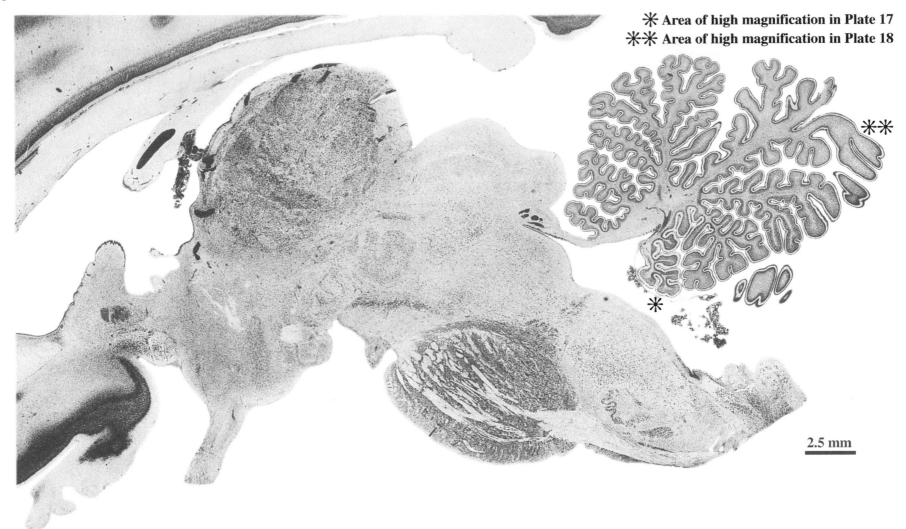

※ **Area of high magnification in Plate 17**
※※ **Area of high magnification in Plate 18**

2.5 mm

PLATE 10A
CR 310 mm, GW 37, Y180-61
Sagittal, Section 941

See the entire Section 941 in
Plates 2A and B.

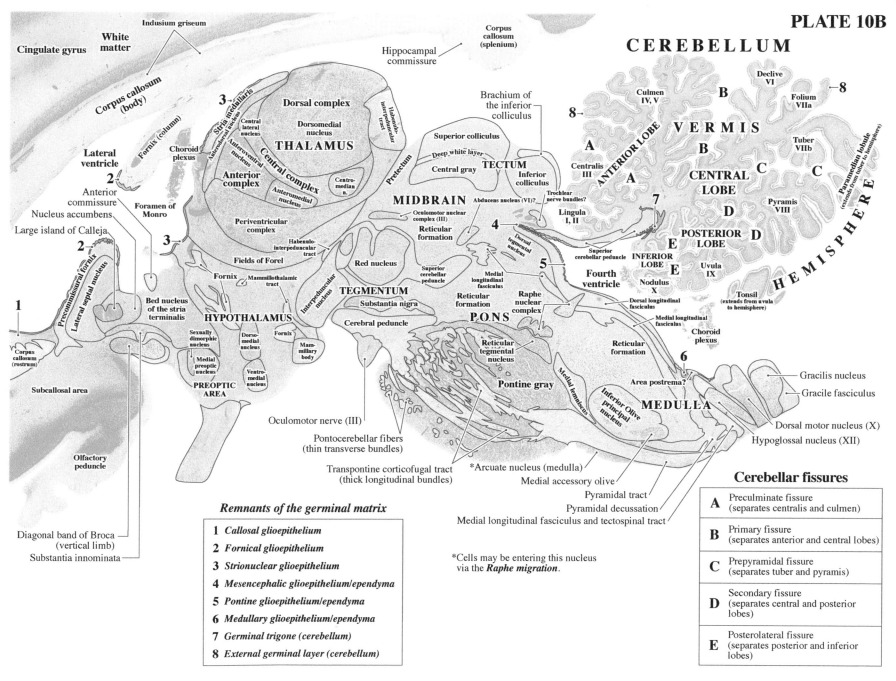

Remnants of the germinal matrix

1 *Callosal glioepithelium*

2 *Fornical glioepithelium*

3 *Strionuclear glioepithelium*

4 *Mesencephalic glioepithelium/ependyma*

5 *Pontine glioepithelium/ependyma*

6 *Medullary glioepithelium/ependyma*

7 *Germinal trigone (cerebellum)*

8 *External germinal layer (cerebellum)*

*Cells may be entering this nucleus
via the *Raphe migration*.

Cerebellar fissures

A	Preculminate fissure (separates centralis and culmen)
B	Primary fissure (separates anterior and central lobes)
C	Prepyramidal fissure (separates tuber and pyramis)
D	Secondary fissure (separates central and posterior lobes)
E	Posterolateral fissure (separates posterior and inferior lobes)

PLATE 11A
CR 310 mm, GW 37, Y180-61
Sagittal, Section 841

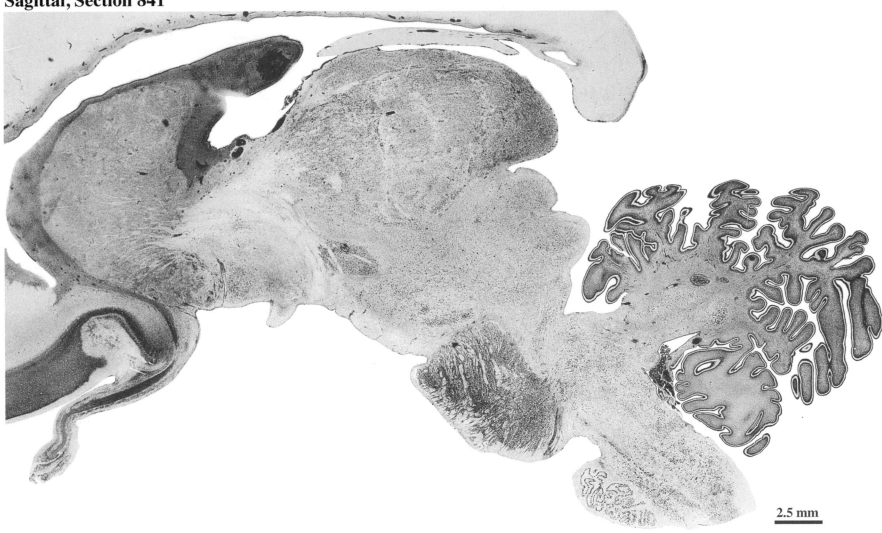

2.5 mm

See the entire Section 841 in
Plates 3A and B.

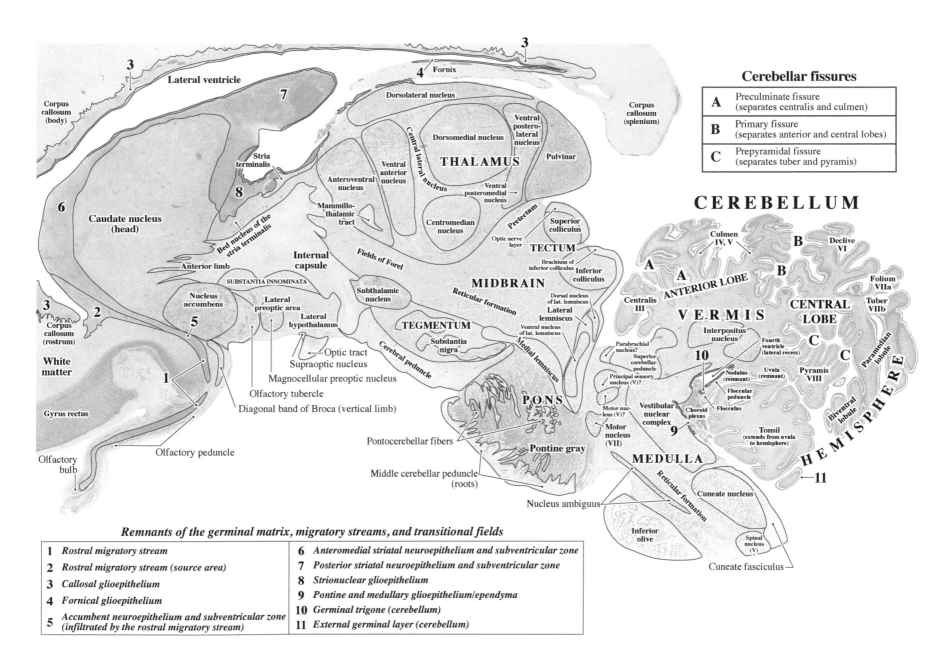

Cerebellar fissures

A	Preculminate fissure (separates centralis and culmen)
B	Primary fissure (separates anterior and central lobes)
C	Prepyramidal fissure (separates tuber and pyramis)

Remnants of the germinal matrix, migratory streams, and transitional fields

1	*Rostral migratory stream*	6	*Anteromedial striatal neuroepithelium and subventricular zone*
2	*Rostral migratory stream (source area)*	7	*Posterior striatal neuroepithelium and subventricular zone*
3	*Callosal glioepithelium*	8	*Strionuclear glioepithelium*
4	*Fornical glioepithelium*	9	*Pontine and medullary glioepithelium/ependyma*
5	*Accumbent neuroepithelium and subventricular zone (infiltrated by the rostral migratory stream)*	10	*Germinal trigone (cerebellum)*
		11	*External germinal layer (cerebellum)*

PLATE 12A
CR 310 mm, GW 37, Y180-61, Sagittal, Section 781

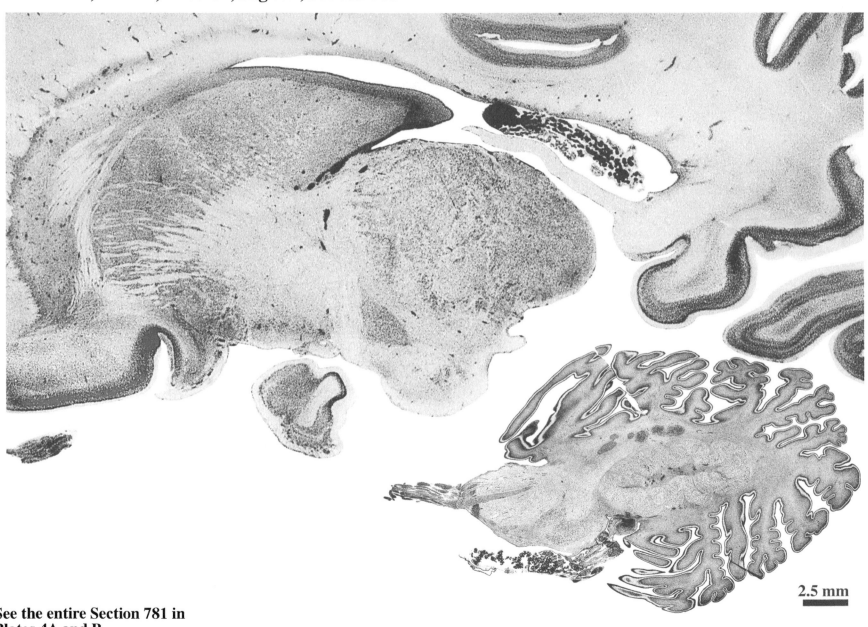

2.5 mm

See the entire Section 781 in
Plates 4A and B.

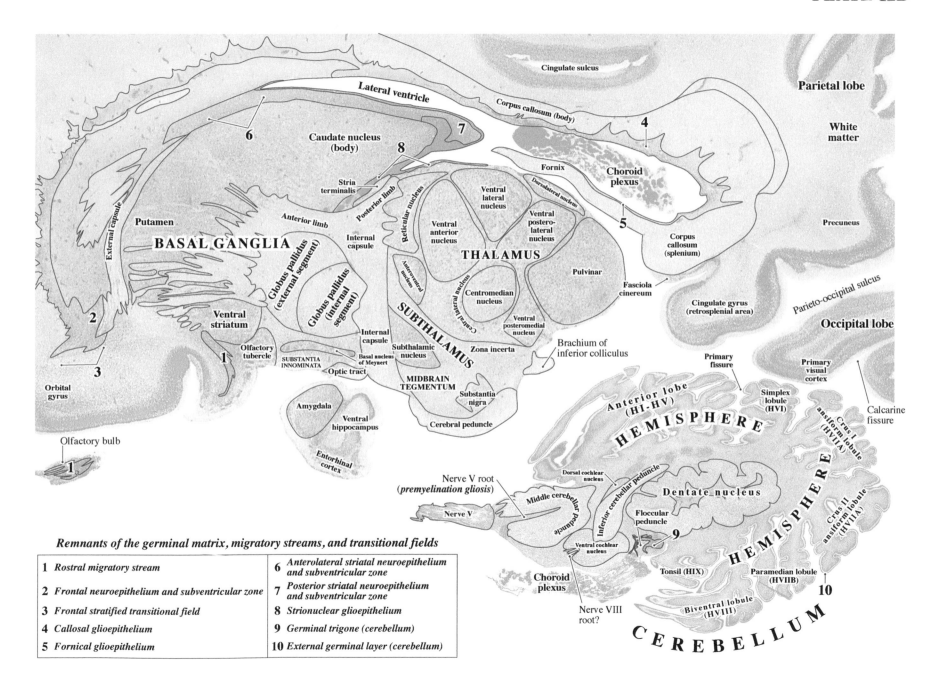

Remnants of the germinal matrix, migratory streams, and transitional fields

1 *Rostral migratory stream*	**6**	*Anterolateral striatal neuroepithelium and subventricular zone*
2 *Frontal neuroepithelium and subventricular zone*	**7**	*Posterior striatal neuroepithelium and subventricular zone*
3 *Frontal stratified transitional field*	**8**	*Strionuclear glioepithelium*
4 *Callosal glioepithelium*	**9**	*Germinal trigone (cerebellum)*
5 *Fornical glioepithelium*	**10**	*External germinal layer (cerebellum)*

PLATE 13A
CR 310 mm, GW m37, Y180-61, Sagittal, Section 721

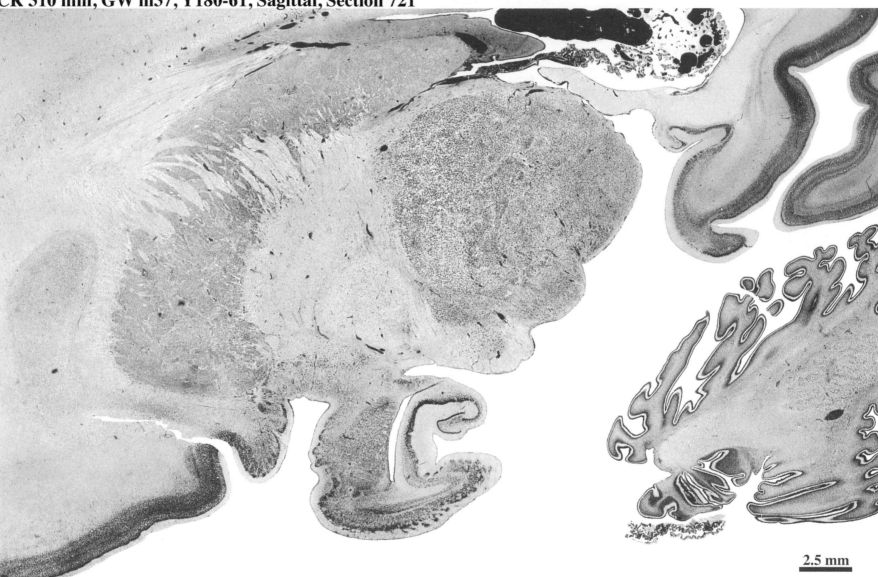

2.5 mm

See the entire Section 721 in
Plates 5A and B.

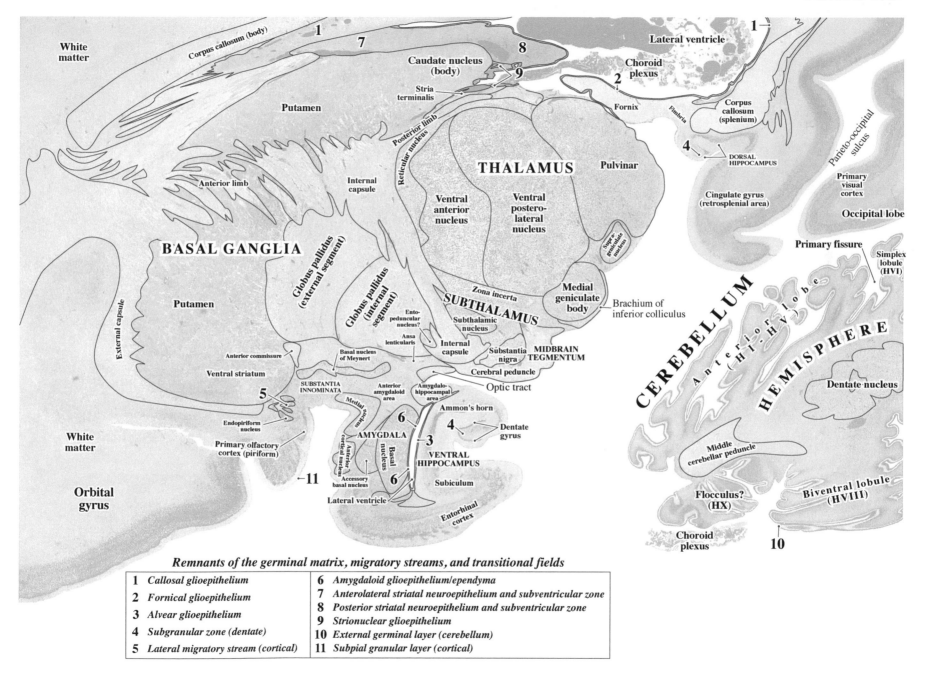

Remnants of the germinal matrix, migratory streams, and transitional fields

1	*Callosal glioepithelium*	6	*Amygdaloid glioepithelium/ependyma*
2	*Fornical glioepithelium*	7	*Anterolateral striatal neuroepithelium and subventricular zone*
3	*Alvear glioepithelium*	8	*Posterior striatal neuroepithelium and subventricular zone*
4	*Subgranular zone (dentate)*	9	*Strionuclear glioepithelium*
5	*Lateral migratory stream (cortical)*	10	*External germinal layer (cerebellum)*
		11	*Subpial granular layer (cortical)*

PLATE 14A
CR 310 mm, GW 37, Y180-61, Sagittal, Section 661

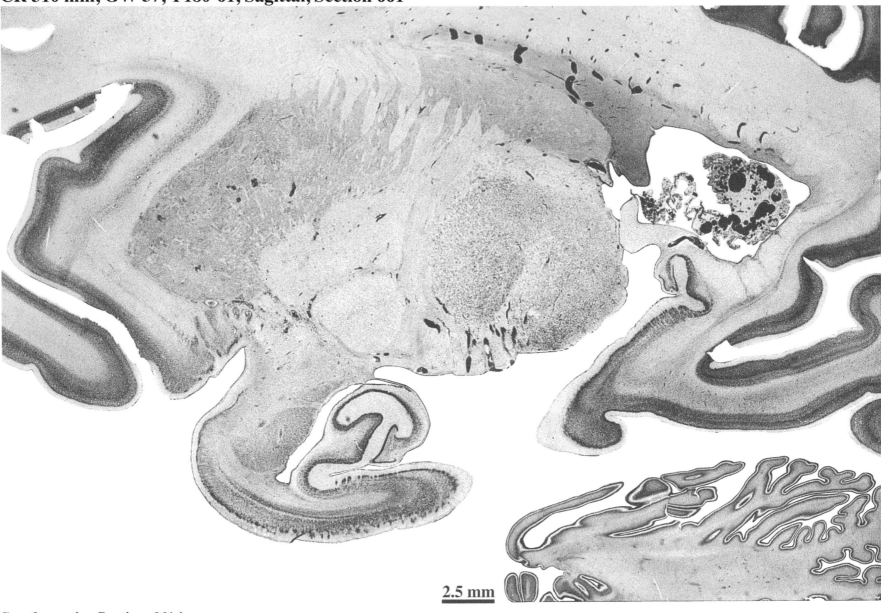

2.5 mm

**See the entire Section 661 in
Plates 6A and B.**

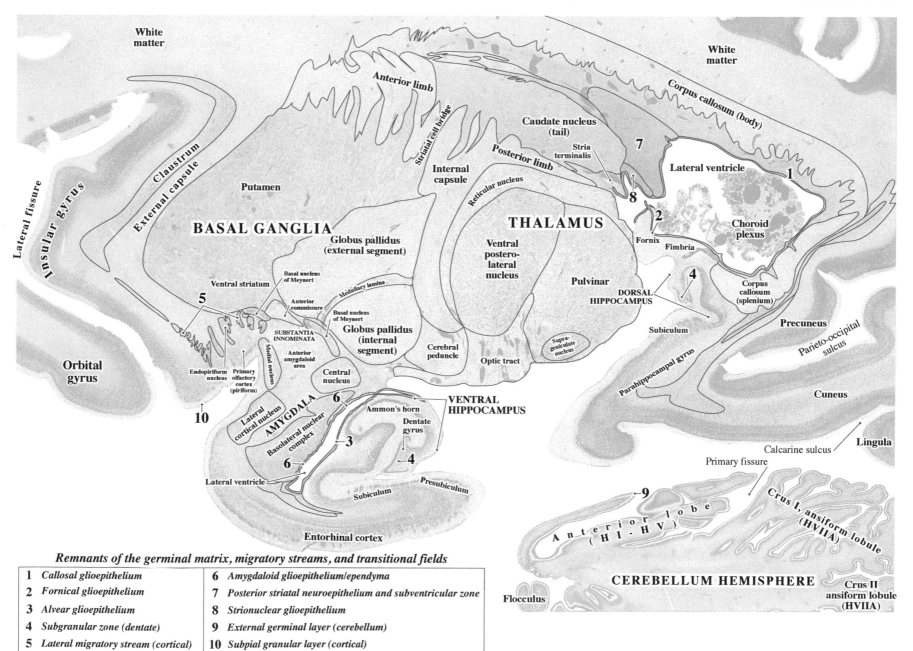

White matter

Anterior limb

White matter

Corpus callosum (body)

Striatal cell bridge

Caudate nucleus (tail)

Posterior limb

Stria terminalis

7

Lateral ventricle

1

Claustrum

External capsule

Internal capsule

Reticular nucleus

8

2

THALAMUS

Choroid plexus

Putamen

BASAL GANGLIA

Globus pallidus (external segment)

Ventral postero-lateral nucleus

Fornix

Fimbria

4

Corpus callosum (splenium)

Ventral striatum

Basal nucleus of Meynert

Medullary lamina

Pulvinar

DORSAL HIPPOCAMPUS

Precuneus

Anterior commissure

Basal nucleus of Meynert

Subiculum

Parieto-occipital sulcus

5

SUBSTANTIA INNOMINATA

Globus pallidus (internal segment)

Cerebral peduncle

Supra-geniculate nucleus

Cuneus

Endopiriform nucleus

Primary olfactory cortex (piriform)

Medial nucleus

Anterior amygdaloid area

Central nucleus

Optic tract

Parahippocampal gyrus

Orbital gyrus

10

Lateral cortical nucleus

AMYGDALA

Basolateral nuclear complex

6

6

Ammon's horn

Dentate gyrus

VENTRAL HIPPOCAMPUS

Lingula

3

4

Lateral ventricle

Subiculum

Presubiculum

Calcarine sulcus

Primary fissure

Entorhinal cortex

9

Anterior (HI - HV) lobe

Crus I, ansiform lobule (HVIIA)

CEREBELLUM HEMISPHERE

Crus II ansiform lobule (HVIIA)

Flocculus

Remnants of the germinal matrix, migratory streams, and transitional fields

1	Callosal glioepithelium	6	Amygdaloid glioepithelium/ependyma
2	Fornical glioepithelium	7	Posterior striatal neuroepithelium and subventricular zone
3	Alvear glioepithelium	8	Strionuclear glioepithelium
4	Subgranular zone (dentate)	9	External germinal layer (cerebellum)
5	Lateral migratory stream (cortical)	10	Subpial granular layer (cortical)

PLATE 15A
CR 310 mm, GW 37, Y180-61, Sagittal, Section 621

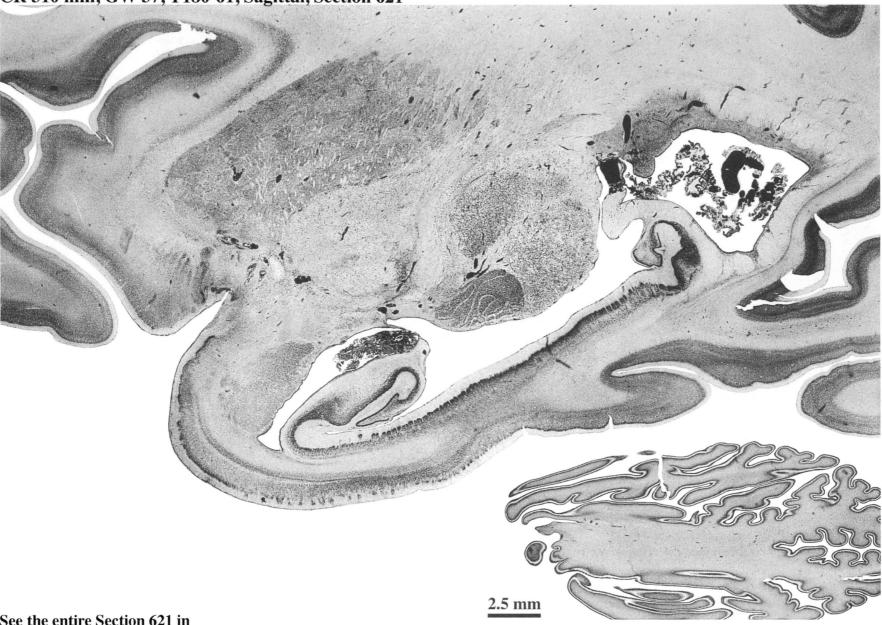

2.5 mm

See the entire Section 621 in
Plates 7A and B.

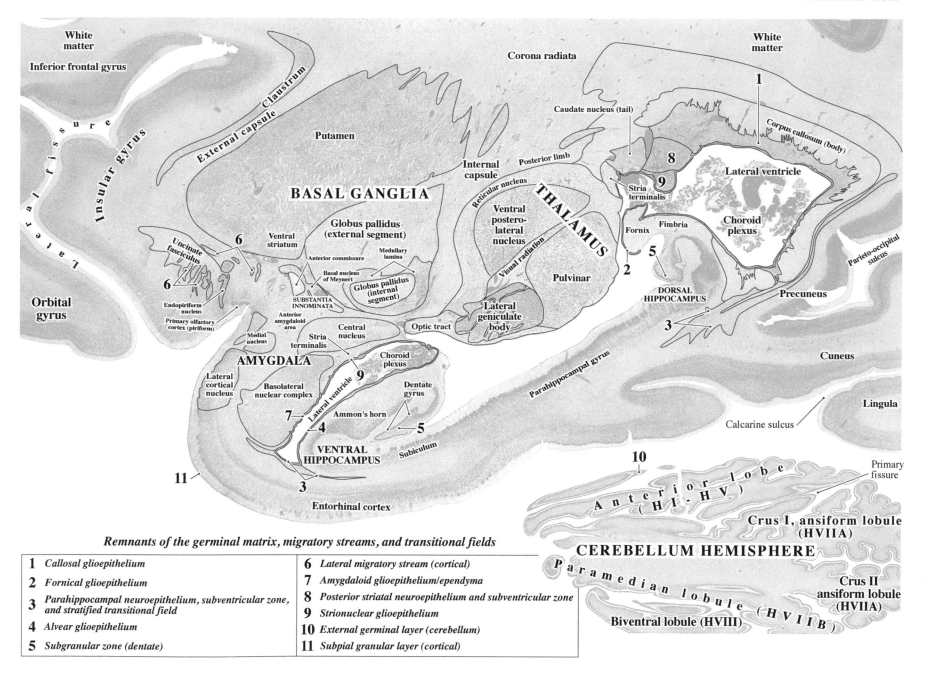

White matter
Inferior frontal gyrus
White matter
Corona radiata
Caudate nucleus (tail)
Corpus callosum (body)
1
External capsule
Claustrum
Putamen
Posterior limb
8
Lateral ventricle
Insular gyrus
Internal capsule
Reticular nucleus
9
Stria terminalis
BASAL GANGLIA
THALAMUS
Ventral postero-lateral nucleus
Fornix
Fimbria
Choroid plexus
Lateral fissure
Globus pallidus (external segment)
Parieto-occipital sulcus
Ventral striatum
Visual radiation
5
Uncinate fasciculus
Anterior commissure
Medullary lamina
Pulvinar
2
Precuneus
6
Basal nucleus of Meynert
DORSAL HIPPOCAMPUS
6
Globus pallidus (internal segment)
SUBSTANTIA INNOMINATA
3
Orbital gyrus
Endopiriform nucleus
Anterior amygdaloid area
Lateral geniculate body
Cuneus
Primary olfactory cortex (piriform)
Central nucleus
Optic tract
Medial nucleus
Stria terminalis
Choroid plexus
Parahippocampal gyrus
Lingula
AMYGDALA
9
Dentate gyrus
Lateral cortical nucleus
Basolateral nuclear complex
Lateral ventricle
Calcarine sulcus
7
Ammon's horn
4
5
VENTRAL HIPPOCAMPUS
Subiculum
10
Primary fissure
11
Entorhinal cortex
Anterior lobe (HI - HV)
Crus I, ansiform lobule (HVIIA)
CEREBELLUM HEMISPHERE
Paramedian lobule (HVIIB)
Crus II ansiform lobule (HVIIA)
Biventral lobule (HVIII)

Remnants of the germinal matrix, migratory streams, and transitional fields

1	*Callosal glioepithelium*	**6**	*Lateral migratory stream (cortical)*
2	*Fornical glioepithelium*	**7**	*Amygdaloid glioepithelium/ependyma*
3	*Parahippocampal neuroepithelium, subventricular zone, and stratified transitional field*	**8**	*Posterior striatal neuroepithelium and subventricular zone*
4	*Alvear glioepithelium*	**9**	*Strionuclear glioepithelium*
5	*Subgranular zone (dentate)*	**10**	*External germinal layer (cerebellum)*
		11	*Subpial granular layer (cortical)*

PLATE 16A, CR 310 mm, GW 37, Y180-61, Sagittal, Section 501

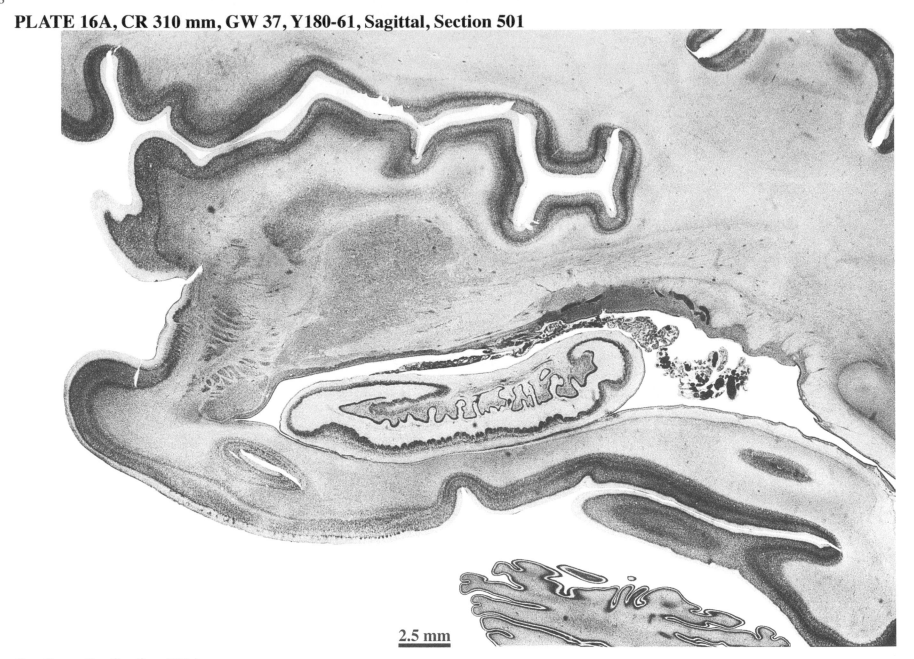

2.5 mm

**See the entire Section 501 in
Plates 8A and B.**

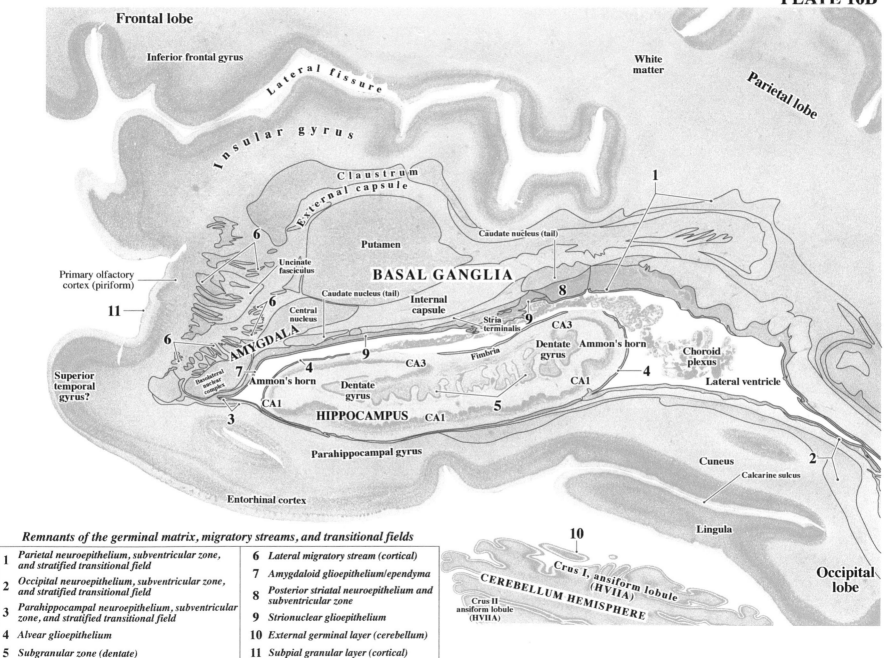

Frontal lobe

Inferior frontal gyrus

Lateral fissure

White matter

Parietal lobe

Insular gyrus

Claustrum
External capsule

6

Putamen

Uncinate fasciculus

Caudate nucleus (tail)

1

Primary olfactory cortex (piriform)

6

Central nucleus

BASAL GANGLIA

Caudate nucleus (tail)

Internal capsule

8

11

Stria terminalis

9

CA3

6

AMYGDALA

9

Fimbria

Dentate gyrus

Ammon's horn

Choroid plexus

Superior temporal gyrus?

Basolateral nuclear complex

7

4

Ammon's horn

CA3

CA1

4

Lateral ventricle

3

CA1

Dentate gyrus

HIPPOCAMPUS   CA1

5

Parahippocampal gyrus

Cuneus

2

Calcarine sulcus

Entorhinal cortex

Lingula

10

Crus I, ansiform lobule (HVIIA)

Occipital lobe

Crus II ansiform lobule (HVIIA)

CEREBELLUM HEMISPHERE

Remnants of the germinal matrix, migratory streams, and transitional fields

1 Parietal neuroepithelium, subventricular zone, and stratified transitional field

2 Occipital neuroepithelium, subventricular zone, and stratified transitional field

3 Parahippocampal neuroepithelium, subventricular zone, and stratified transitional field

4 Alvear glioepithelium

5 Subgranular zone (dentate)

6 Lateral migratory stream (cortical)

7 Amygdaloid glioepithelium/ependyma

8 Posterior striatal neuroepithelium and subventricular zone

9 Strionuclear glioepithelium

10 External germinal layer (cerebellum)

11 Subpial granular layer (cortical)

PLATE 17, CR 310 mm
GW 37, Y180-61, Sagittal, Section 941

CEREBELLUM, VERMIS, INFERIOR LOBE, LOBULE X: NODULUS
(A relatively mature region of the cerebellar cortex)

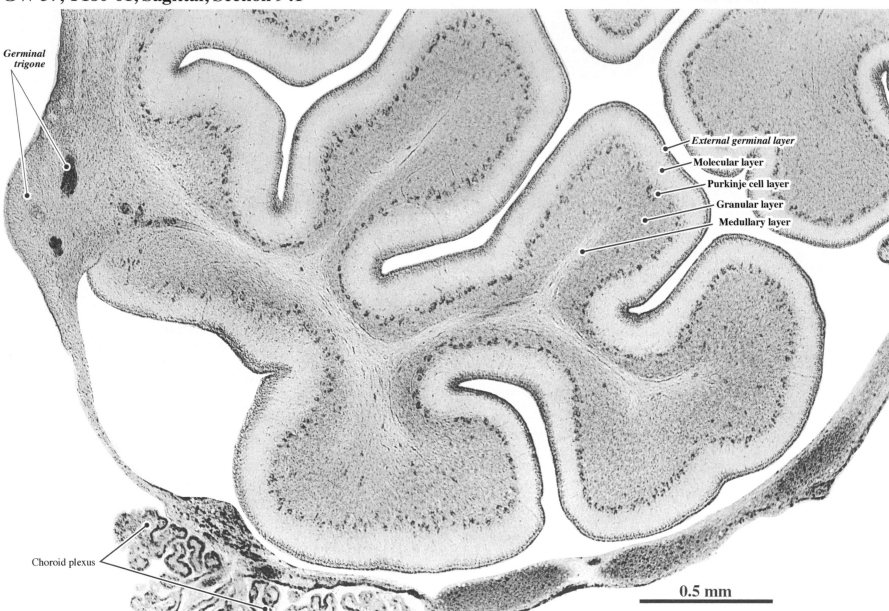

Germinal trigone

External germinal layer
Molecular layer
Purkinje cell layer
Granular layer
Medullary layer

Choroid plexus

0.5 mm

See the entire Section 941 in Plates 2A and B.
See larger view of the entire cerebellum in Plates 10A and B.

PLATE 18
CR 310 mm, GW 37, Y180-61

Sagittal, Section 941
CEREBELLUM, HEMISPHERE, PARAMEDIAN LOBULE
(A relatively immature region of the cerebellar cortex)

Sagittal, Section 781,**CEREBELLUM, HEMISPHERE, CORTEX**
(very high magnification)

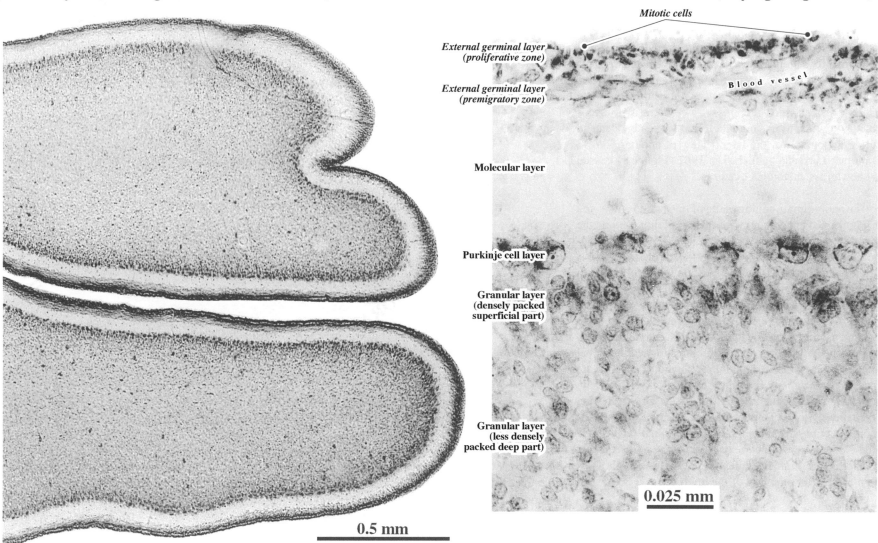

Mitotic cells

*External germinal layer
(proliferative zone)*

*External germinal layer
(premigratory zone)*

Blood vessel

Molecular layer

Purkinje cell layer

Granular layer
(densely packed
superficial part)

Granular layer
(less densely
packed deep part)

0.025 mm

0.5 mm

See the entire Section 941 in Plates 2A and B.
See larger view of the entire cerebellum in Plates 10A and B.

See the entire Section 781 in Plates 4A and B.
See larger view of the entire cerebellum in Plates 12A and B.

PART III: Y301-62
CR 320 mm (GW 37)
Horizontal

This specimen is case number B-301-62 (Perinatal RPSL) in the Yakovlev Collection. A male infant was stillborn after a premature separation of the placenta. This brain is classified as a Normative Control in the Yakovlev Collection (Haleem, 1990). It was cut in the horizontal plane in 35-μm thick sections. Since there is no available photograph of this brain before it was embedded and cut, a photograph of the lateral view of another GW 37 brain that Larroche published in 1967 (**Figure 5**) is used to show gross anatomical features.

The approximate cutting plane of this brain is indicated in **Figure 6** (facing page) with lines superimposed on the GW 37 brain from the Larroche (1967) series. The anterior part of each section (on the left in all photographs) is slightly dorsal to the posterior part. In terms of the medial-lateral plane of sectioning, the part of each section near the bottom of the page shows structures cut slightly dorsal to those on the other side of the midline. As in all other specimens, the sections chosen for illustration are spaced closer together to show small structures in the diencephalon, midbrain, pons, and medulla. Illustrated sections are spaced farther apart when they contain only large brain structures, such as the cerebral cortex, basal ganglia, and cerebellum. Low-magnification photographs of 19 Nissl-stained sections are shown in **Plates 19–32**. **Plates 33–37** are below the cerebral cortex and show the brainstem and cerebellum at high magnification. **Plates 38–49** show high-magnification views of the brain core.

In the cortical regions of the telencephalon, remnants of the germinal matrices are present in all lobes of the cerebral cortex where the **neuroepithelium/subventricular zone** may still be generating neocortical interneurons. Remnants of migrating and sojourning neurons and/or

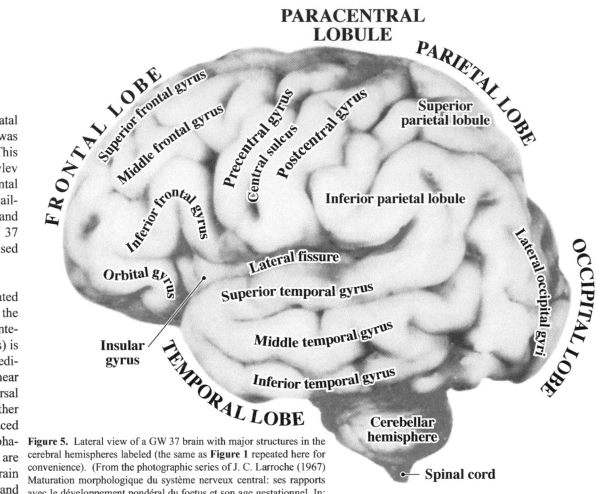

Figure 5. Lateral view of a GW 37 brain with major structures in the cerebral hemispheres labeled (the same as **Figure 1** repeated here for convenience). (From the photographic series of J. C. Larroche (1967) Maturation morphologique du système nerveux central: ses rapports avec le développement pondéral du foetus et son age gestationnel. In: *Regional Development of the Brain in Early Life*, A. Minkowski (ed.), London: Blackwell, page 248.)

glia are visible in the dwindling **stratified transitional fields** in all lobes of the cerebral cortex, more prominent in parietal, temporal, and occipital lobes. Because of its plane of sectioning, this specimen shows the most complete view of the occipital lobes at GW 37.

Many neurons, glia, and their mitotic precursor cells are still migrating through the olfactory peduncle toward the olfactory bulb (**rostral migratory stream**) from a presumed source area in the germinal matrix at the junction between the cerebral cortex, striatum, and nucleus accumbens. Within the lateral parts of the cerebral cortex, streams of neurons and glia are still in the **lateral migratory stream**. That stream percolates through the claustrum, endopiriform nucleus, external capsule, and uncinate fasciculus,

GW 37 HORIZONTAL SECTION PLANES

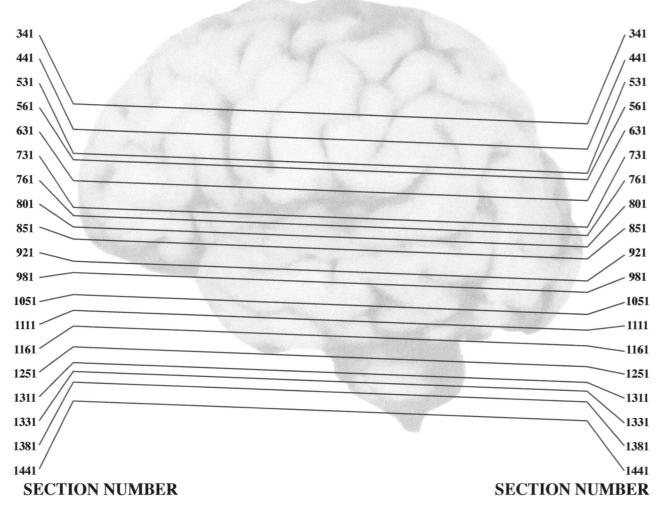

341
441
531
561
631
731
761
801
851
921
981
1051
1111
1161
1251
1311
1331
1381
1441

341
441
531
561
631
731
761
801
851
921
981
1051
1111
1161
1251
1311
1331
1381
1441

SECTION NUMBER

SECTION NUMBER

Figure 6. Lateral view of the same GW 37 brain shown in **Figure 4** with the approximate locations and cutting angle of the sections of Y301-62. (From the photographic series of J. C. Larroche (1967) Maturation morphologique du système nerveux central: ses rapports avec le développement pondéral du foetus et son age gestationnel. In: *Regional Development of the Brain in Early Life*, A. Minkowski (ed.), London: Blackwell, page 248.)

and cells appear to be heading toward the insular cortex, primary olfactory cortex, temporal cortex, and basolateral parts of the amygdaloid complex.

In the basal ganglia, there is a prominent *neuroepithelium/subventricular zone* overlying the striatum and nucleus accumbens where neurons are being generated. Another region of active neurogenesis in the telencephalon is the *subgranular zone* in the hilus of the dentate gyrus that is generating granule cells. Other structures in the telencephalon, such as the septum, fornix, and Ammon's horn have only a thin, darkly staining layer at the ventricle, and these are presumed to be generating glia, cells of the choroid plexus, and the ependymal lining of the ventricle.

Most of the structures in the diencephalon appear to be settled and are maturing, and the third ventricle is lined by a thin *glioepithelium/ependyma*. In the midbrain and anterior pons, there is a slightly thicker and more convoluted *glioepithelium/ependyma* lining the posterior cerebral aqueduct and anterior fourth ventricle. The posterior pons and entire medulla have a thin *glioepithelium/ependyma* lining the rest of the fourth ventricle.

The *external germinal layer* is prominent over the entire surface of the cerebellar cortex and is still producing basket, stellate, and granule cells. The *germinal trigone* is still visible at the base of the nodulus and along the floccular peduncle; choroid plexus cells and glia may still be originating here.

PLATE 19A
CR 320 mm
GW 37
Y301-62
Horizontal
Section 341

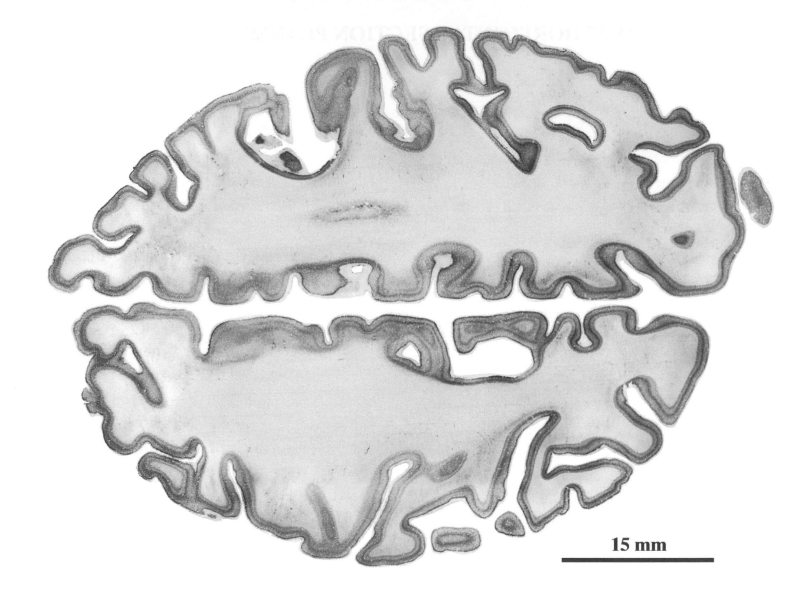

15 mm

Remnants of the
germinal matrix and
transitional fields

1 *Subpial granular layer*

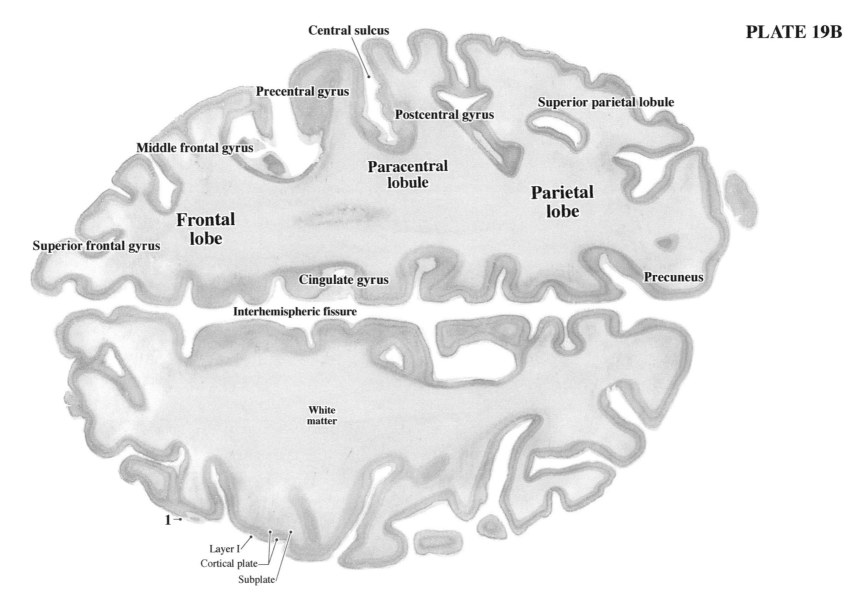

Central sulcus

Precentral gyrus

Postcentral gyrus

Superior parietal lobule

Middle frontal gyrus

Paracentral lobule

Parietal lobe

Frontal lobe

Superior frontal gyrus

Cingulate gyrus

Precuneus

Interhemispheric fissure

White matter

1

Layer I

Cortical plate

Subplate

**PLATE 20A
CR 320 mm
GW 37
Y301-62
Horizontal
Section 441**

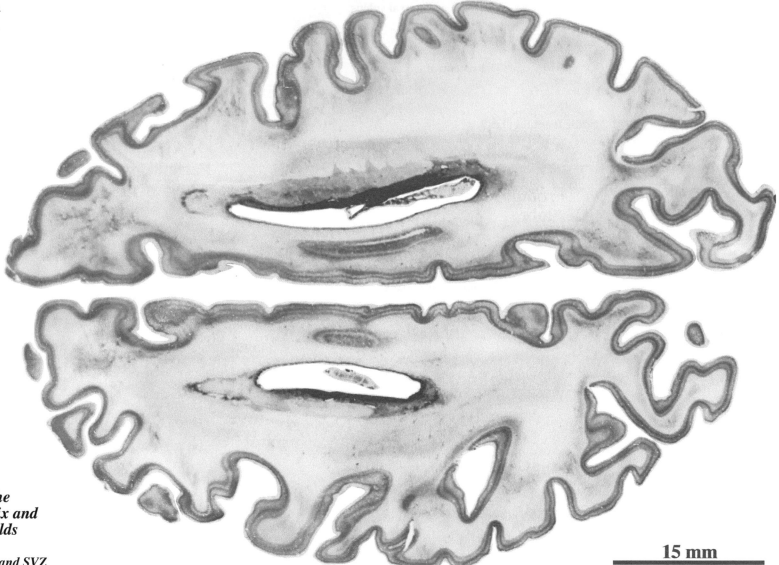

*Remnants of the
germinal matrix and
transitional fields*

1 *Frontal NEP and SVZ*

2 *Frontal STF*

3 *Callosal GEP*

4 *Parietal NEP and SVZ*

5 *Parietal STF*

6 *Paracentral STF*

7 *Anterolateral striatal NEP and SVZ*

8 *Posterior striatal NEP and SVZ*

9 *Strionuclear GEP*

10 *Subpial granular layer*

GEP - Glioepithelium
NEP - Neuroepithelium
STF - Stratified transitional field
SVZ - Subventricular zone

15 mm

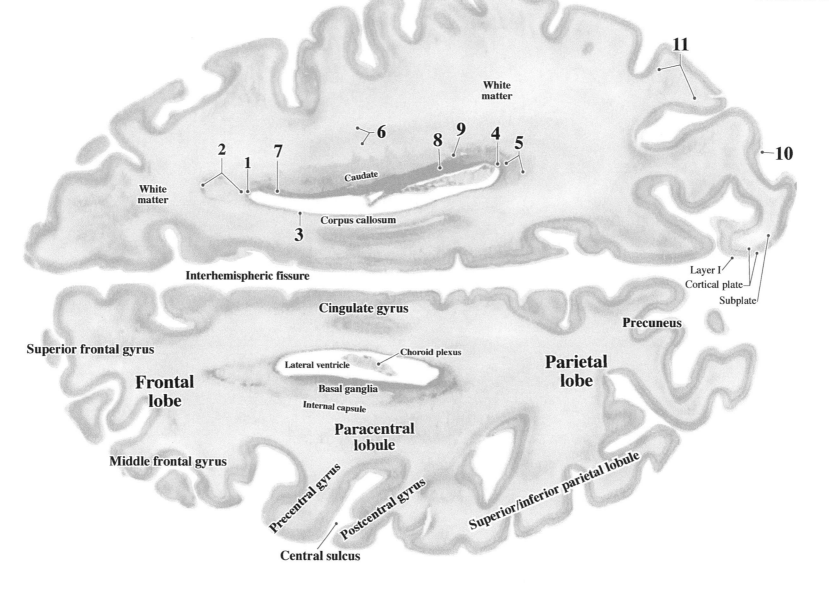

11

White
matter

6

8 **9** **4** **5**

2 **7**

1

White
matter

Caudate

3 Corpus callosum

10

Layer I
Cortical plate
Subplate

Interhemispheric fissure

Cingulate gyrus

Precuneus

Superior frontal gyrus

Choroid plexus

Lateral ventricle

**Frontal
lobe**

**Parietal
lobe**

Basal ganglia

Internal capsule

**Paracentral
lobule**

Middle frontal gyrus

Precentral gyrus

Postcentral gyrus

Superior/inferior parietal lobule

Central sulcus

PLATE 21A
CR 320 mm
GW 37
Y301-62
Horizontal
Section 531

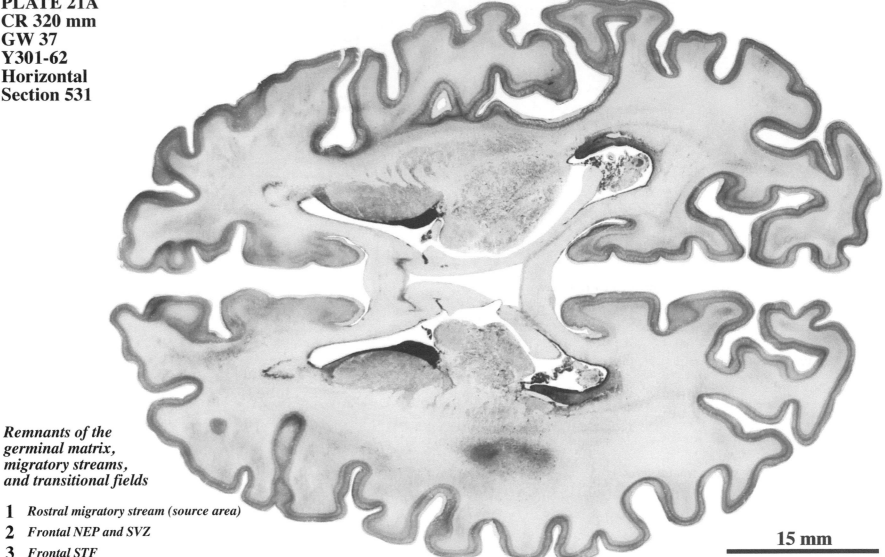

Remnants of the
germinal matrix,
migratory streams,
and transitional fields

1 *Rostral migratory stream (source area)*

2 *Frontal NEP and SVZ*

3 *Frontal STF*

4 *Callosal GEP*

5 *Callosal sling*

6 *Fornical GEP*

7 *Parietal NEP and SVZ*

8 *Parietal STF*

9 *Posterior striatal NvEP and SVZ*

10 *Anterolateral striatal NEP and SVZ*

11 *Anteromedial striatal NEP and SVZ*

12 *Strionuclear GEP*

GEP - Glioepithelium
NEP - Neuroepithelium
STF - Stratified transitional field
SVZ - Subventricular zone

15 mm

See detail of brain core
in Plates 38A and B.

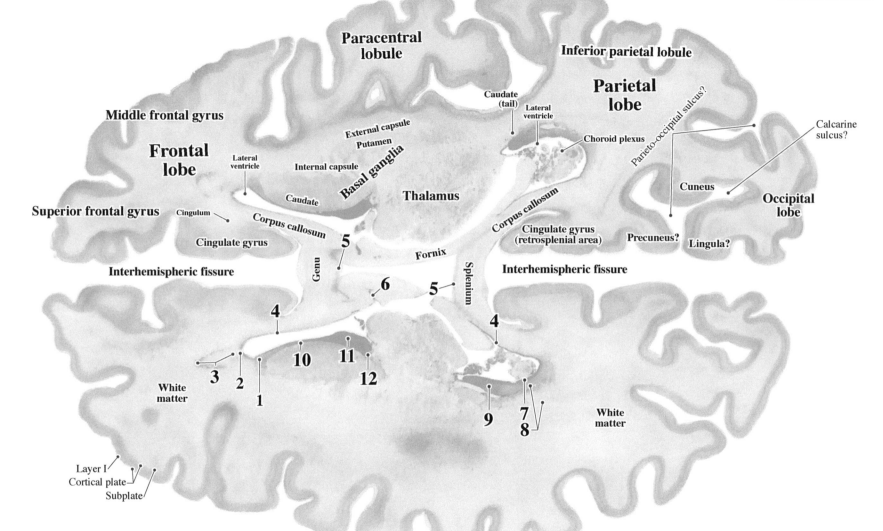

Paracentral lobule

Inferior parietal lobule

Middle frontal gyrus

Parietal lobe

Caudate (tail)

Lateral ventricle

Choroid plexus

Frontal lobe

External capsule

Putamen

Parieto-occipital sulcus?

Calcarine sulcus?

Lateral ventricle

Internal capsule

Basal ganglia

Thalamus

Cuneus

Superior frontal gyrus

Cingulum

Caudate

Corpus callosum

Corpus callosum

Cingulate gyrus (retrosplenial area)

Occipital lobe

Cingulate gyrus

5

Fornix

Precuneus?

Lingula?

Genu

Interhemispheric fissure

Splenium

Interhemispheric fissure

6

5

4

4

10

11

3

2

12

White matter

1

9

7

8

White matter

Layer I

Cortical plate

Subplate

PLATE 22A
CR 320 mm
GW 37
Y301-62
Horizontal
Section 561

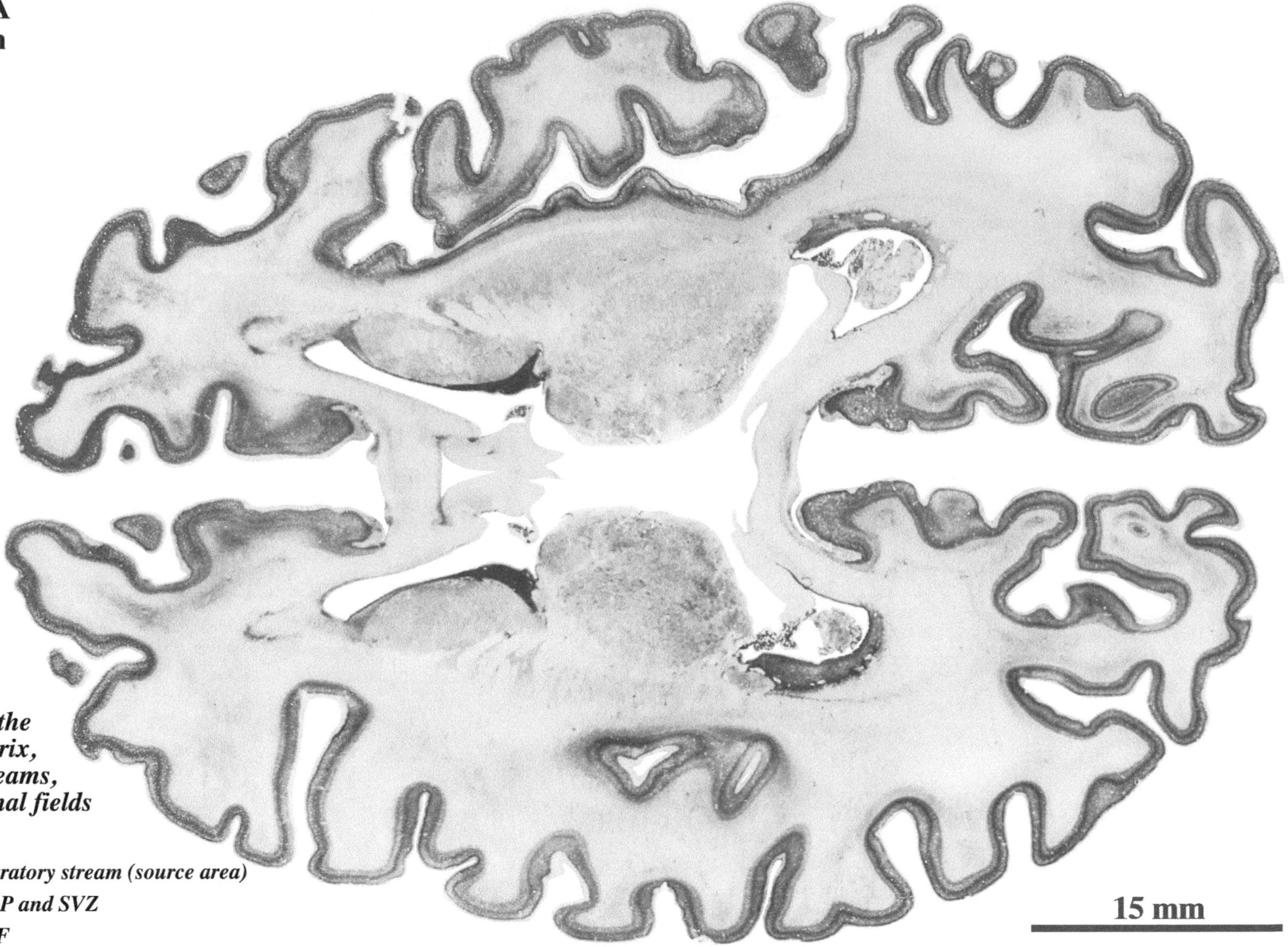

Remnants of the
germinal matrix,
migratory streams,
and transitional fields

1 *Rostral migratory stream (source area)*
2 *Frontal NEP and SVZ*
3 *Frontal STF*
4 *Callosal GEP*
5 *Callosal sling*
6 *Fornical GEP*
7 *Parietal/occipital NEP and SVZ*
8 *Parietal/occipital STF*

9 *Posterior striatal NEP and SVZ*
10 *Anterolateral striatal NEP and SVZ*
11 *Anteromedial striatal NEP and SVZ*
12 *Strionuclear GEP*

GEP - Glioepithelium
NEP - Neuroepithelium
STF - Stratified transitional field
SVZ - Subventricular zone

15 mm

See detail of brain core
in Plates 39A and B.

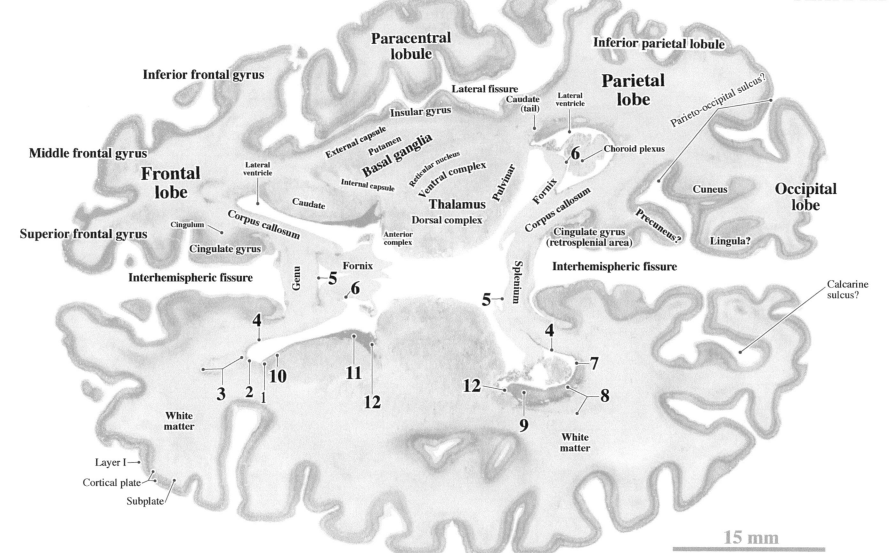

Paracentral
lobule

Inferior parietal lobule

Inferior frontal gyrus

Lateral fissure

Parietal
lobe

Insular gyrus

Caudate
(tail)

Lateral
ventricle

Parieto-occipital sulcus?

Middle frontal gyrus

External capsule

Putamen

Basal ganglia

Choroid plexus

6

Lateral
ventricle

Frontal
lobe

Reticular nucleus

Ventral complex

Cuneus

Occipital
lobe

Internal capsule

Pulvinar

Fornix

Caudate

Thalamus

Corpus callosum

Superior frontal gyrus

Cingulum

Corpus callosum

Dorsal complex

Precuneus?

Cingulate gyrus

Anterior
complex

Cingulate gyrus
(retrosplenial area)

Lingula?

Fornix

Interhemispheric fissure

Genu

5

6

Splenium

Interhemispheric fissure

Calcarine
sulcus?

5

4

4

7

3

2

1

10

11

12

12

8

White
matter

9

White
matter

Layer I

Cortical plate

Subplate

15 mm

PLATE 23A
CR 320 mm
GW 37
Y301-62
Horizontal
Section 631

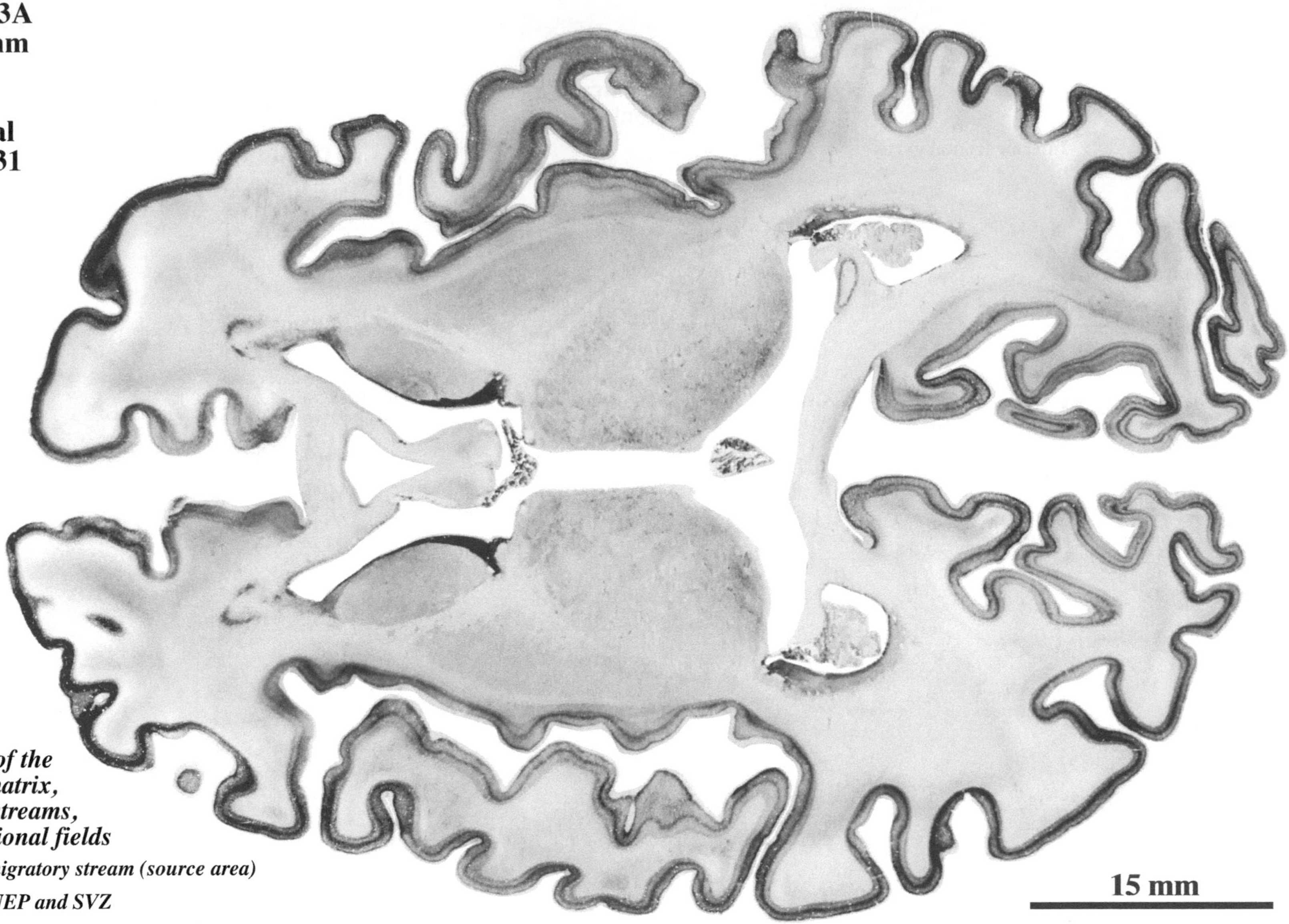

Remnants of the
germinal matrix,
migratory streams,
and transitional fields

1 *Rostral migratory stream (source area)*

2 *Frontal NEP and SVZ*

3 *Frontal STF*

4 *Callosal GEP*

5 *Callosal sling*

6 *Fornical GEP*

7 *Parietal/occipital NEP and SVZ*

8 *Parietal/occipital STF*

9 *Posterior striatal NEP and SVZ*

10 *Anterolateral striatal NEP and SVZ*

11 *Anteromedial striatal NEP and SVZ*

12 *Strionuclear GEP*

GEP - Glioepithelium
NEP - Neuroepithelium
STF - Stratified transitional field
SVZ - Subventricular zone

15 mm

See detail of brain core
in Plates 40A and B.

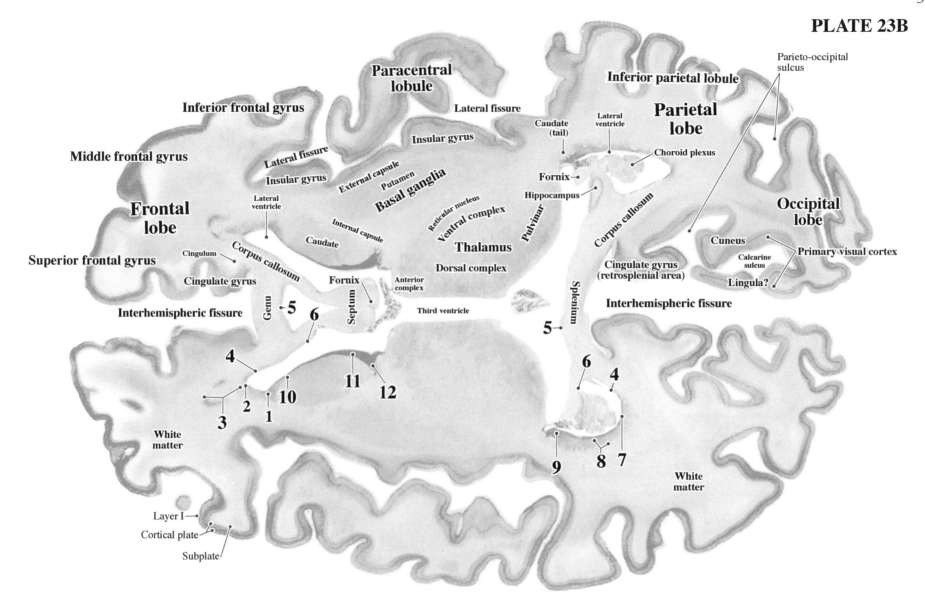

Parieto-occipital sulcus

Paracentral lobule

Inferior parietal lobule

Inferior frontal gyrus

Lateral fissure

Insular gyrus

Caudate (tail)

Lateral ventricle

Parietal lobe

Middle frontal gyrus

Lateral fissure

External capsule

Putamen

Basal ganglia

Choroid plexus

Insular gyrus

Insular gyrus

Fornix

Frontal lobe

Lateral ventricle

Reticular nucleus

Ventral complex

Hippocampus

Occipital lobe

Internal capsule

Pulvinar

Corpus callosum

Superior frontal gyrus

Cingulum

Corpus callosum

Caudate

Thalamus

Cuneus

Primary visual cortex

Cingulate gyrus (retrosplenial area)

Calcarine sulcus

Dorsal complex

Cingulate gyrus

Fornix

Anterior complex

Lingula?

Interhemispheric fissure

Genu

5

6

Septum

Third ventricle

Splenium

Interhemispheric fissure

5

4

6

4

11

10

12

2

1

3

9

8

7

White matter

White matter

Layer I

Cortical plate

Subplate

54

PLATE 24A
CR 320 mm
GW 37
Y301-62
Horizontal
Section 731

See detail of brain core
in Plates 41A and B.

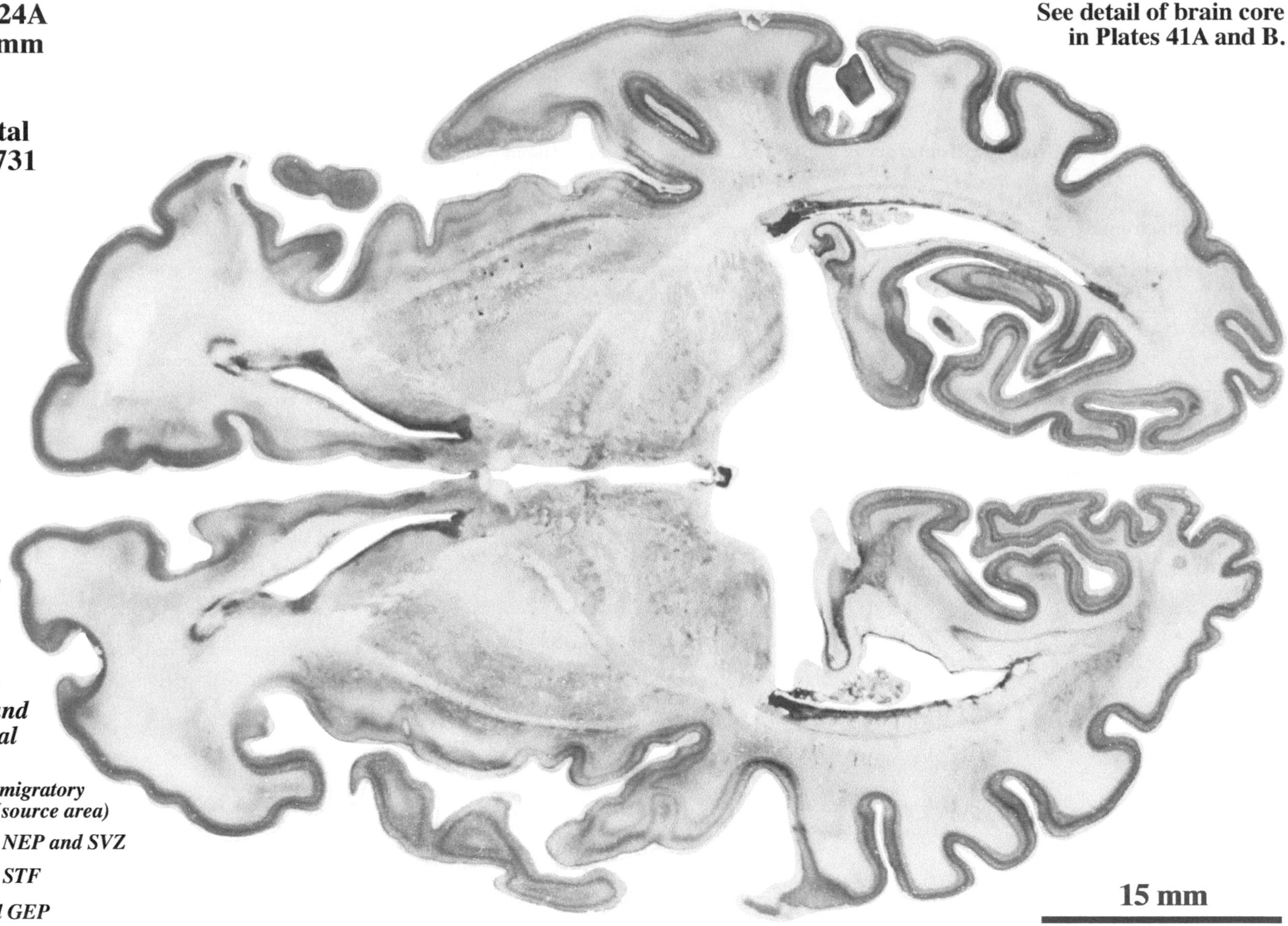

Remnants
of the
germinal
matrix,
migratory
streams, and
transitional
fields

1 *Rostral migratory stream (source area)*

2 *Frontal NEP and SVZ*

3 *Frontal STF*

4 *Callosal GEP*

5 *Fornical GEP*

6 *Occipital NEP and SVZ*

7 *Occipital STF*

8 *Parietal/temporal NEP and SVZ*

9   *Parietal/temporal STF*

10 *Alvear GEP*

11 *Posterior striatal NEP and SVZ*

12 *Anterolateral striatal NEP and SVZ*

13 *Anteromedial striatal NEP and SVZ*

14 *Strionuclear GEP*

15 *Septal G/EP*

15 mm

GEP - Glioepithelium
G/EP - Glioepithelium/ependyma
NEP - Neuroepithelium
STF - Stratified transitional field
SVZ - Subventricular zone

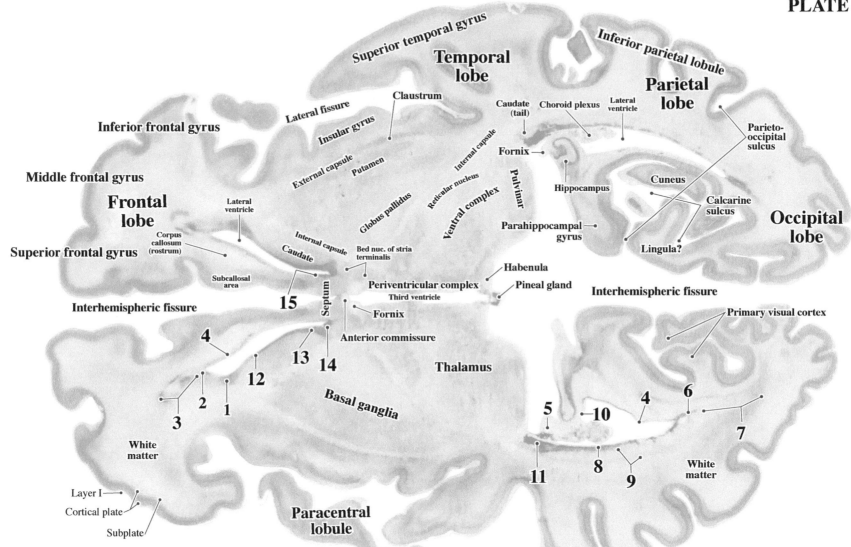

Superior temporal gyrus

Inferior parietal lobule

Temporal lobe

Parietal lobe

Lateral fissure

Claustrum

Caudate (tail)

Choroid plexus

Lateral ventricle

Parieto-occipital sulcus

Inferior frontal gyrus

Insular gyrus

Middle frontal gyrus

External capsule

Putamen

Internal capsule

Fornix

Cuneus

Calcarine sulcus

Frontal lobe

Lateral ventricle

Globus pallidus

Reticular nucleus

Pulvinar

Hippocampus

Occipital lobe

Superior frontal gyrus

Corpus callosum (rostrum)

Internal capsule

Ventral complex

Parahippocampal gyrus

Lingula?

Caudate

Bed nuc. of stria terminalis

Subcallosal area

Habenula

Pineal gland

Interhemispheric fissure

15

Septum

Periventricular complex

Third ventricle

Interhemispheric fissure

Primary visual cortex

4

13

14

Fornix

Anterior commissure

Thalamus

12

2

1

Basal ganglia

5

10

4

6

3

7

White matter

11

8

9

White matter

Layer I

Cortical plate

Subplate

Paracentral lobule

PLATE 25A
CR 320 mm
GW 37
Y301-62
Horizontal
Section 761

See detail of brain core
in Plates 42A and B.

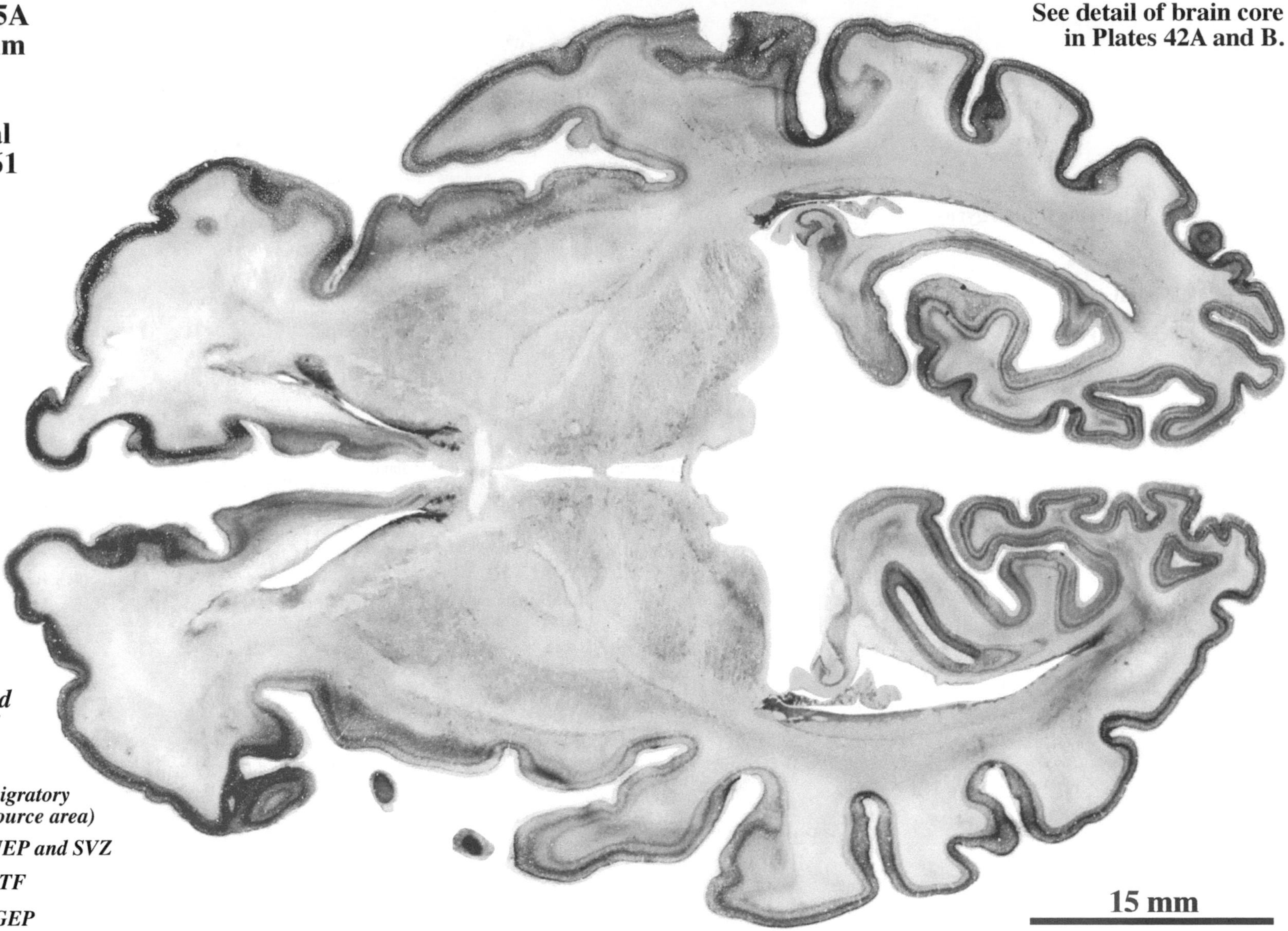

15 mm

Remnants
of the
germinal
matrix,
migratory
streams, and
transitional
fields

1 *Rostral migratory*
  stream (source area)

2 *Frontal NEP and SVZ*

3 *Frontal STF*

4 *Callosal GEP*

5 *Fornical GEP*

6 *Occipital NEP and SVZ*

7 *Occipital STF*

8 *Parietal/temporal NEP and SVZ*

9 *Parietal/temporal STF*

10 *Alvear GEP*

11 *Subgranular zone*

12 *Posterior striatal NEP and SVZ*

13 *Anterolateral striatal NEP and SVZ*

14 *Anteromedial striatal NEP and SVZ*

15 *Strionuclear GEP*

GEP - Glioepithelium
NEP - Neuroepithelium
STF - Stratified transitional field
SVZ - Subventricular zone

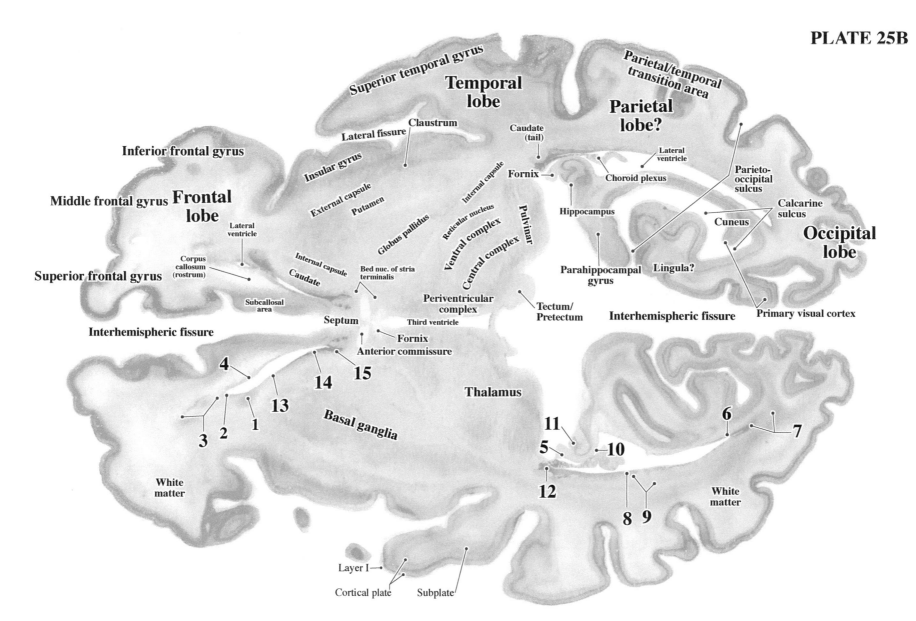

Superior temporal gyrus

Temporal lobe

Parietal/temporal transition area

Parietal lobe?

Occipital lobe

Inferior frontal gyrus

Lateral fissure

Claustrum

Caudate (tail)

Lateral ventricle

Parieto-occipital sulcus

Middle frontal gyrus **Frontal lobe**

Insular gyrus

External capsule

Putamen

Internal capsule

Fornix

Choroid plexus

Calcarine sulcus

Lateral ventricle

Globus pallidus

Reticular nucleus

Hippocampus

Cuneus

Corpus callosum (rostrum)

Internal capsule

Bed nuc. of stria terminalis

Ventral complex

Central complex

Pulvinar

Parahippocampal gyrus

Lingula?

Primary visual cortex

Superior frontal gyrus

Subcallosal area

Caudate

Periventricular complex

Tectum/ Pretectum

Interhemispheric fissure

Septum

Third ventricle

Interhemispheric fissure

Fornix

Anterior commissure

4

14

15

Thalamus

13

Basal ganglia

11

6

2

1

3

5

10

7

White matter

12

White matter

8 **9**

Layer I

Cortical plate

Subplate

PLATE 26A
CR 320 mm
GW 37
Y301-62
Horizontal
Section 801

See detail of brain core
in Plates 43A and B.

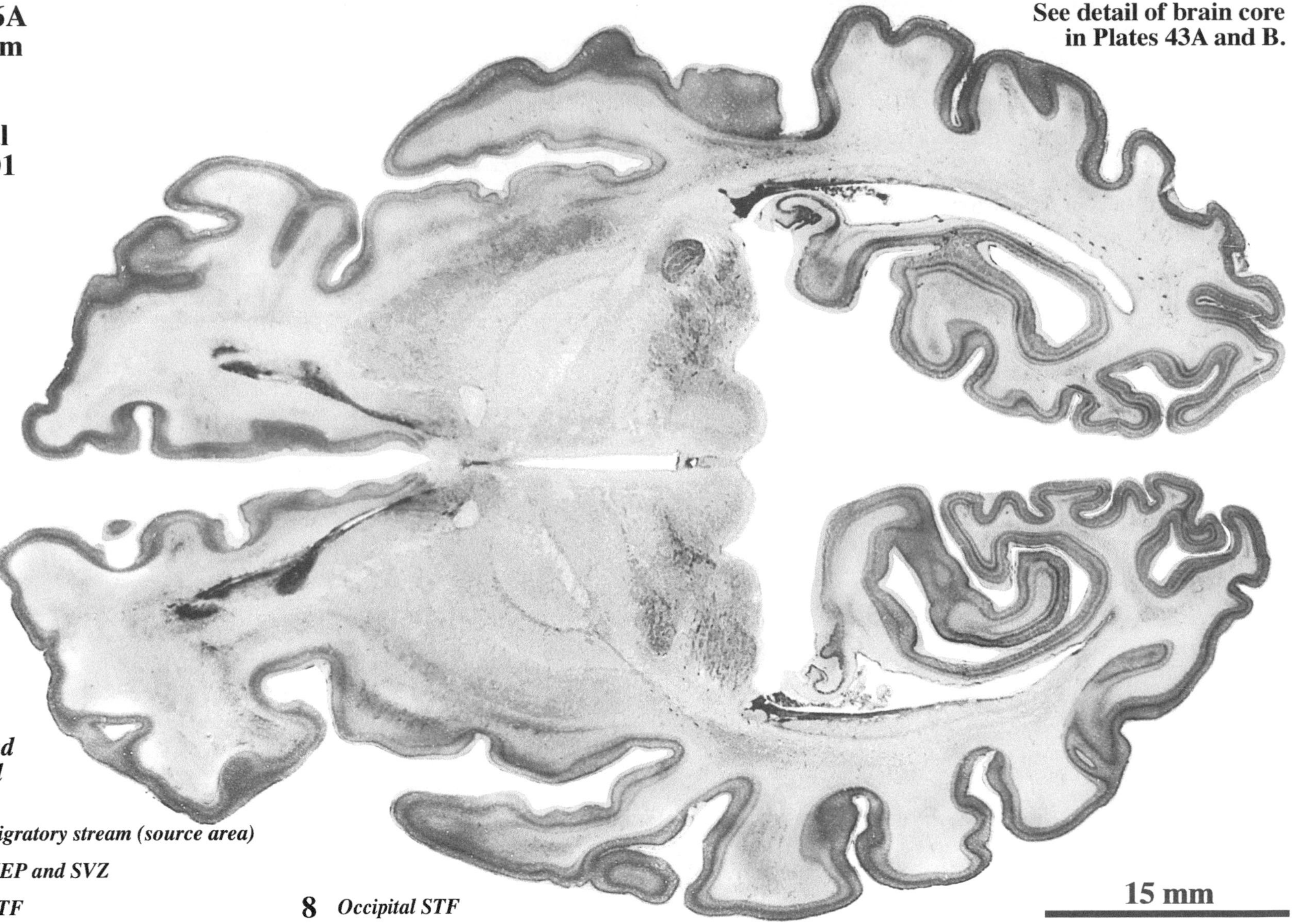

Remnants
of the
germinal
matrix,
migratory
streams, and
transitional
fields

1 *Rostral migratory stream (source area)*

2 *Frontal NEP and SVZ*

3 *Frontal STF*

4 *Callosal GEP*

5 *Fornical GEP*

6 *Parahippocampal NEP, SVZ, and STF*

7 *Occipital NEP and SVZ*

8 *Occipital STF*

9 *Temporal NEP and SVZ*

10 *Temporal STF*

11 *Alvear GEP*

12 *Subgranular zone (dentate)*

13 *Lateral migratory stream (cortical)*

14 *Posterior striatal NEP and SVZ*

15 *Accumbent NEP and SVZ*
(infiltrated by the rostral migratory stream)

15 mm

GEP - Glioepithelium
NEP - Neuroepithelium
STF - Stratified transitional field
SVZ - Subventricular zone

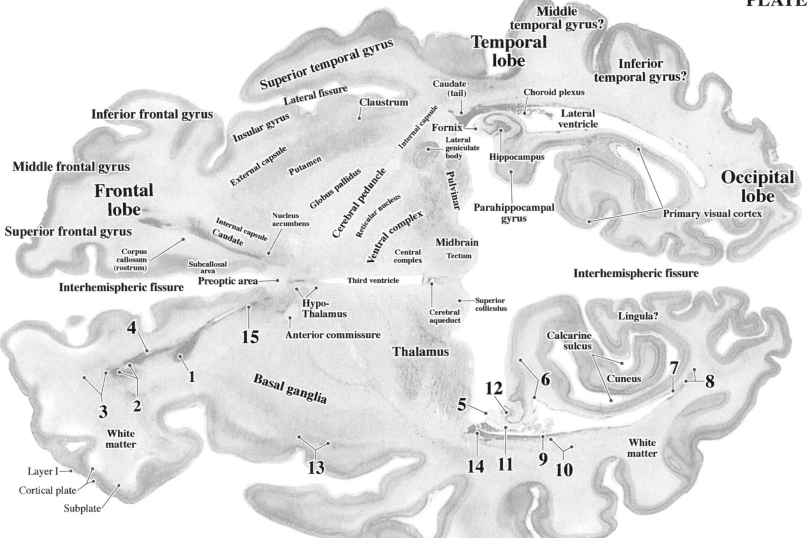

Middle
temporal gyrus?

**Temporal
lobe**

Inferior
temporal gyrus?

Superior temporal gyrus

Lateral fissure

Caudate
(tail)

Choroid plexus

Inferior frontal gyrus

Insular gyrus

Claustrum

Lateral
ventricle

Internal capsule

Fornix

Middle frontal gyrus

External capsule

Putamen

Lateral
geniculate
body

Hippocampus

**Occipital
lobe**

**Frontal
lobe**

Globus pallidus

Cerebral peduncle

Pulvinar

Reticular nucleus

Superior frontal gyrus

Nucleus
accumbens

Ventral complex

Parahippocampal
gyrus

Primary visual cortex

Internal capsule
Caudate

Corpus
callosum
(rostrum)

Subcallosal
area

Central
complex

Midbrain
Tectum

Interhemispheric fissure

Interhemispheric fissure

Preoptic area

Third ventricle

Hypo-
Thalamus

Cerebral
aqueduct

Superior
colliculus

4

15

Anterior commissure

Thalamus

Calcarine
sulcus

Lingula?

1

Basal ganglia

12

6

Cuneus

7

8

3   **2**

5

14   **11**

9

10

White
matter

Layer I

13

White
matter

Cortical plate

Subplate

PLATE 27A
CR 320 mm
GW 37
Y301-62
Horizontal
Section 851

See detail of brain core
in Plates 44A and B.

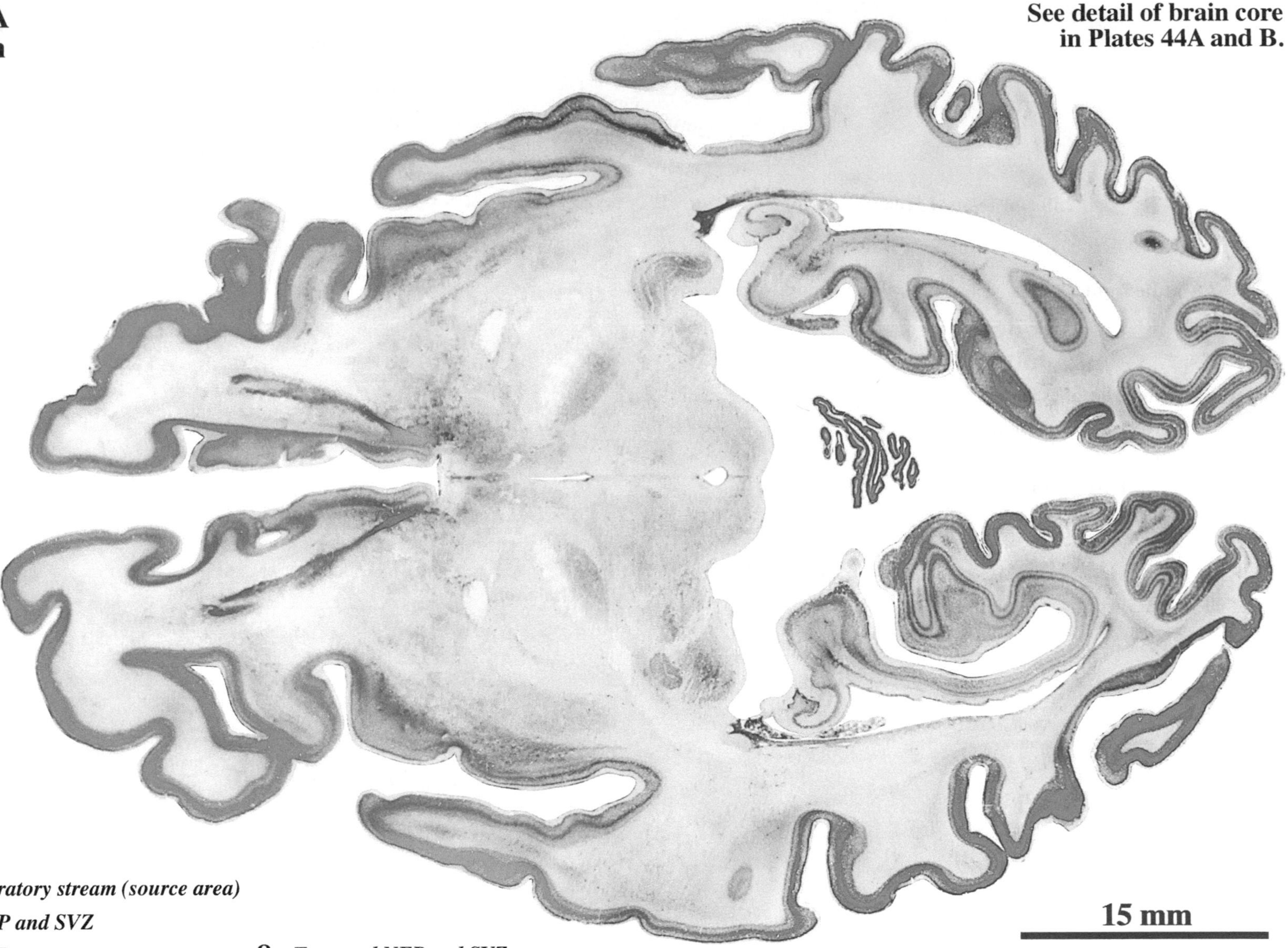

15 mm

Remnants
of the
germinal
matrix,
migratory
streams, and
transitional
fields

1 *Rostral migratory stream (source area)*

2 *Frontal NEP and SVZ*

3 *Frontal STF*

4 *Fornical GEP*

5 *Parahippocampal NEP, SVZ, and STF*

6 *Occipital NEP and SVZ*

7 *Occipital STF*

8 *Temporal NEP and SVZ*

9 *Temporal STF*

10 *Alvear GEP*

11 *Subgranular zone (dentate)*

12 *Lateral migratory stream (cortical)*

13 *Posterior striatal NEP and SVZ*

14 *Accumbent NEP and SVZ*
(infiltrated by the rostral migratory stream)

15 *External germinal layer (cerebellum)*

GEP - Glioepithelium
NEP - Neuroepithelium
STF - Stratified transitional field
SVZ - Subventricular zone

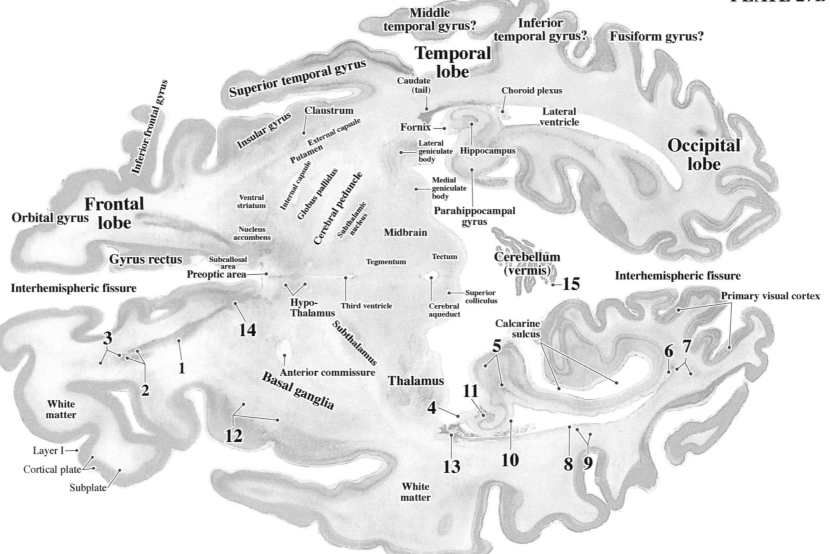

Middle temporal gyrus?

Inferior temporal gyrus?

Fusiform gyrus?

Temporal lobe

Superior temporal gyrus

Inferior frontal gyrus

Caudate (tail)

Choroid plexus

Lateral ventricle

Occipital lobe

Insular gyrus

Claustrum

Fornix

External capsule

Lateral geniculate body

Hippocampus

Putamen

Internal capsule

Globus pallidus

Cerebral peduncle

Medial geniculate body

Frontal lobe

Ventral striatum

Subthalamic nucleus

Parahippocampal gyrus

Orbital gyrus

Nucleus accumbens

Midbrain

Cerebellum (vermis)

Interhemispheric fissure

Gyrus rectus

Subcallosal area

Tegmentum

Tectum

15

Preoptic area

Interhemispheric fissure

Third ventricle

Superior colliculus

Primary visual cortex

Hypo-Thalamus

Cerebral aqueduct

Calcarine sulcus

14

Subthalamus

3

5

6 7

Anterior commissure

1

Basal ganglia

11

2

Thalamus

4

White matter

12

10

8 9

Layer I

13

Cortical plate

Subplate

White matter

PLATE 28A
CR 320 mm
GW 37
Y301-62
Horizontal
Section 921

See detail of brain core
in Plates 45A and B.

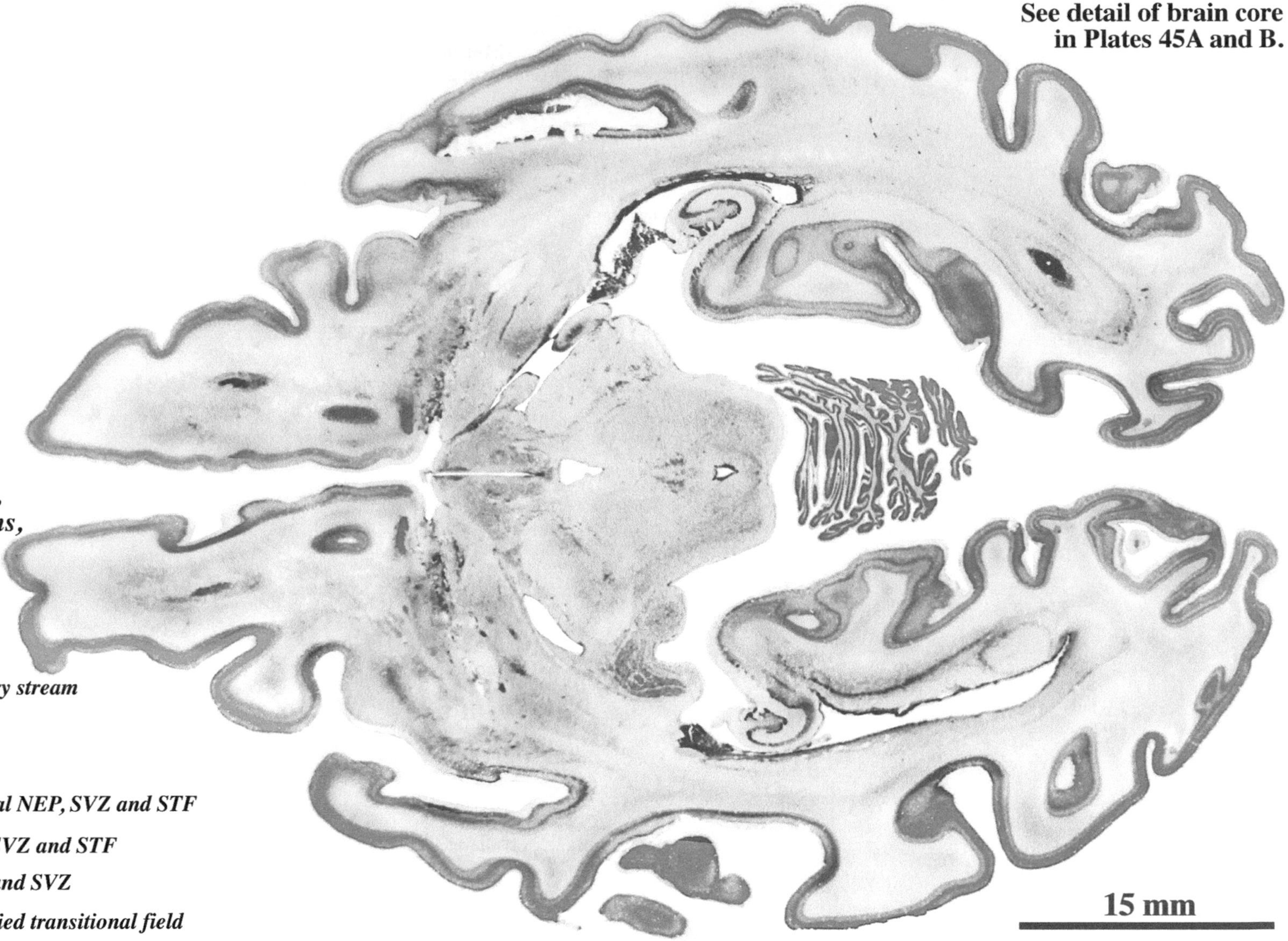

Remnants of the
germinal matrix,
migratory streams,
and
transitional
fields

1 *Rostral migratory stream*

2 *Frontal STF*

3 *Fornical GEP*

4 *Parahippocampal NEP, SVZ and STF*

5 *Occipital NEP, SVZ and STF*

6 *Temporal NEP and SVZ*

7 *Temporal stratified transitional field*

8 *Alvear glioepithelium*

9 *Subgranular zone (dentate)*

10 *Lateral migratory stream (cortical)*

11 *Posterior striatal NEP and SVZ*

12 *External germinal layer (cerebellum)*

GEP - Glioepithelium
NEP - Neuroepithelium
STF - Stratified transitional field
SVZ - Subventricular zone

15 mm

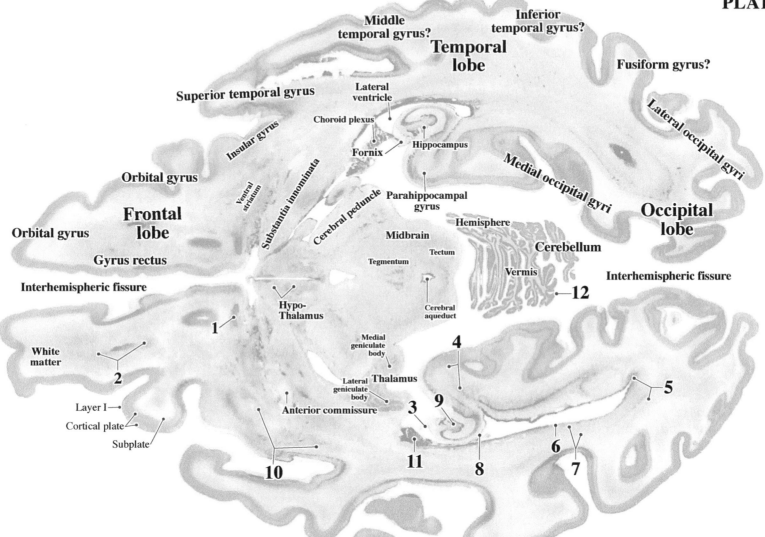

Middle
temporal gyrus?

Inferior
temporal gyrus?

**Temporal
lobe**

Fusiform gyrus?

Superior temporal gyrus

Lateral
ventricle

Lateral occipital gyri

Insular gyrus

Choroid plexus

Hippocampus

Fornix

Medial occipital gyri

Orbital gyrus

Ventral striatum

Substantia innominata

Cerebral peduncle

Parahippocampal
gyrus

**Occipital
lobe**

**Frontal
lobe**

Orbital gyrus

Hemisphere

Midbrain

Cerebellum

Gyrus rectus

Tegmentum

Tectum

Vermis

Interhemispheric fissure

Cerebral
aqueduct

Interhemispheric fissure

Hypo-
Thalamus

1

12

White
matter

Medial
geniculate
body

4

2

Lateral
geniculate
body

Thalamus

9

5

Layer I

3

Cortical plate

Anterior commissure

Subplate

6

10

11

8

7

PLATE 29A
CR 320 mm
GW 37
Y301-62
Horizontal
Section 981

See detail of brain core
in Plates 46A and B.

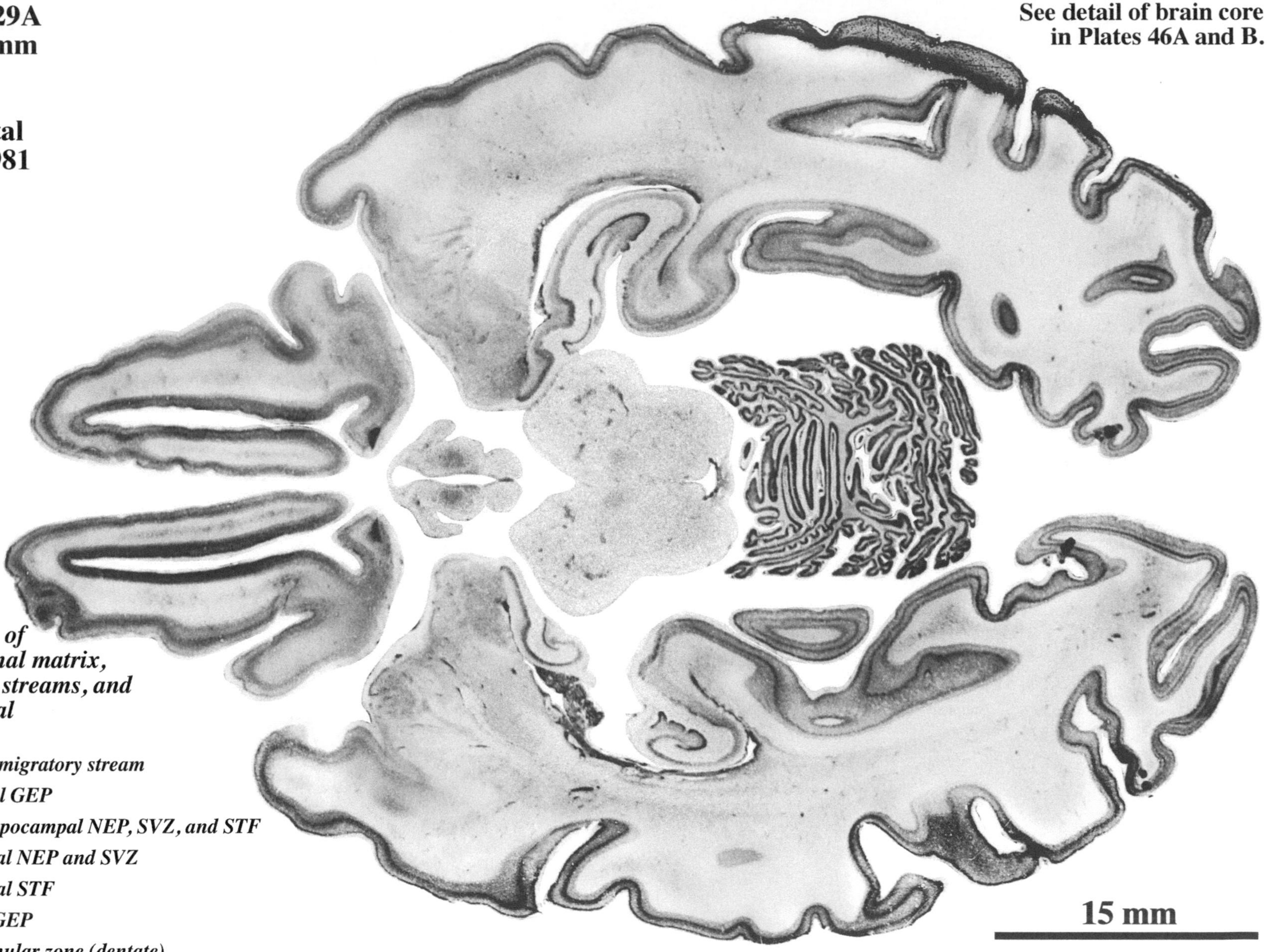

Remnants of
the germinal matrix,
migratory streams, and
transitional
fields

1 *Rostral migratory stream*

2 *Fornical GEP*

3 *Parahippocampal NEP, SVZ, and STF*

4 *Temporal NEP and SVZ*

5 *Temporal STF*

6 *Alvear GEP*

7 *Subgranular zone (dentate)*

8 *Lateral migratory stream (cortical)*

9 *Amygdaloid G/EP*

10 *External germinal layer (cerebellum)*

GEP - Glioepithelium
G/EP - Glioepithelium/ependyma
NEP - Neuroepithelium
STF - Stratified transitional field
SVZ - Subventricular zone

15 mm

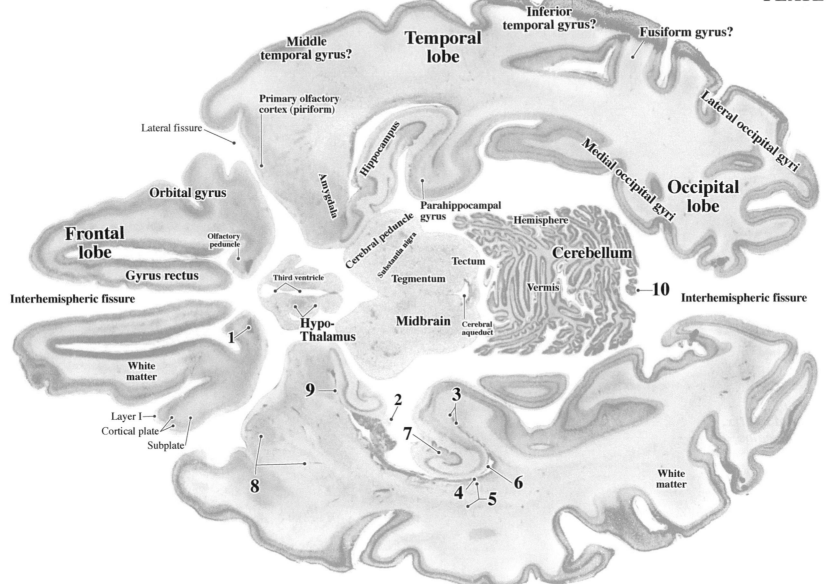

Temporal lobe

Inferior temporal gyrus?

Fusiform gyrus?

Middle temporal gyrus?

Primary olfactory cortex (piriform)

Lateral fissure

Hippocampus

Lateral occipital gyri

Medial occipital gyri

Occipital lobe

Orbital gyrus

Amygdala

Parahippocampal gyrus

Cerebral peduncle

Hemisphere

Cerebellum

Olfactory peduncle

Substantia nigra

Tectum

Frontal lobe

Gyrus rectus

Third ventricle

Tegmentum

Vermis

10

Interhemispheric fissure

Hypo-Thalamus

Midbrain

Cerebral aqueduct

Interhemispheric fissure

1

White matter

9

2

3

Layer I

Cortical plate

7

Subplate

8

4

5

6

White matter

**PLATE 30A
CR 320 mm
GW 37
Y301-62
Horizontal
Section 1051**

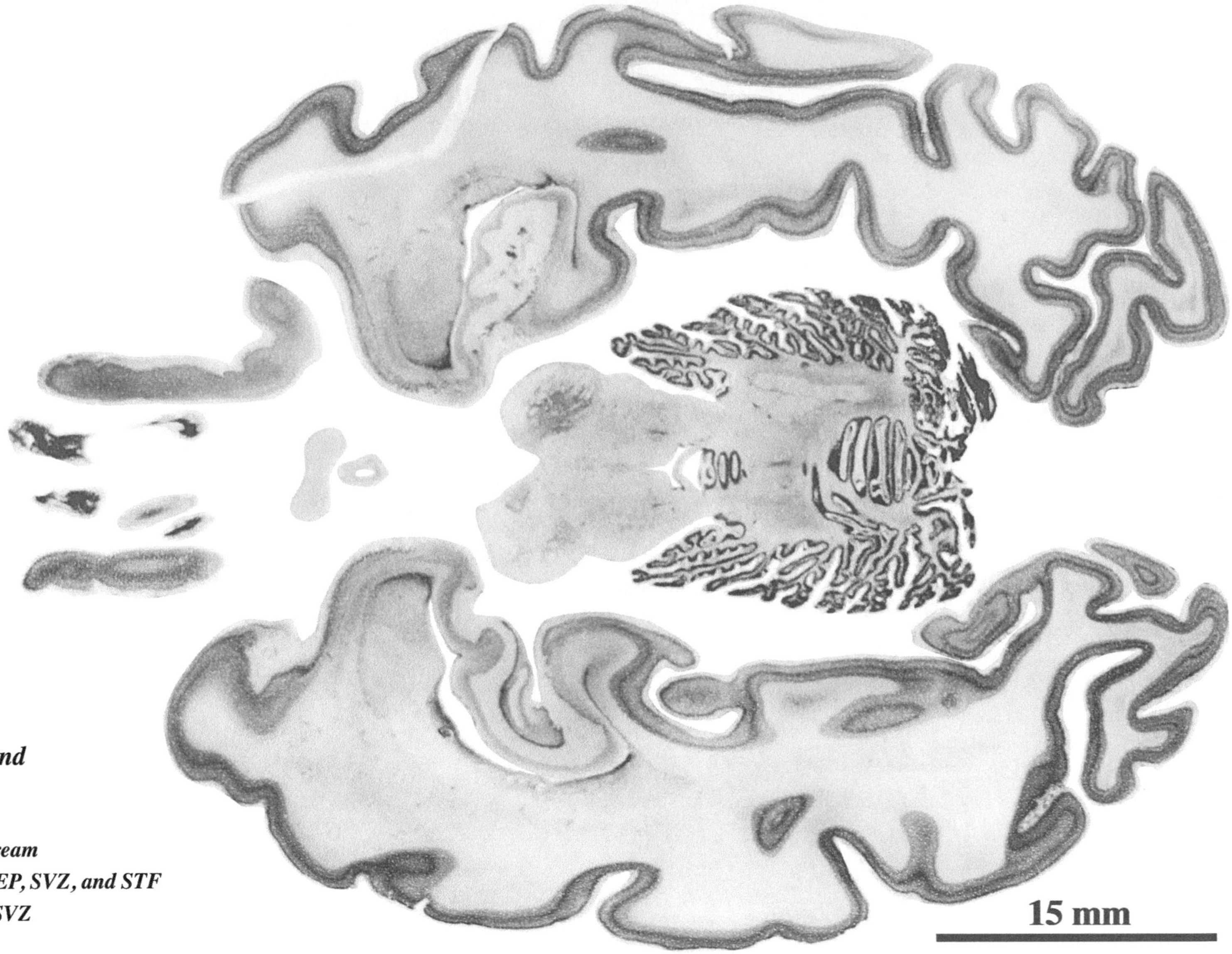

*Remnants of
the germinal matrix,
migratory streams, and
transitional
fields*

1 *Rostral migratory stream*
2 *Parahippocampal NEP, SVZ, and STF*
3 *Temporal NEP and SVZ*
4 *Temporal STF*
5 *Alvear GEP*
6 *Subgranular zone (dentate)*
7 *Lateral migratory stream (cortical)*
8 *Amygdaloid G/EP*
9 *External germinal layer (cerebellum)*

*GEP - Glioepithelium
G/EP - Glioepithelium/ependyma
NEP - Neuroepithelium
STF - Stratified transitional field
SVZ - Subventricular zone*

15 mm

**See detail of brain core and
cerebellum in Plates 47A and B.**

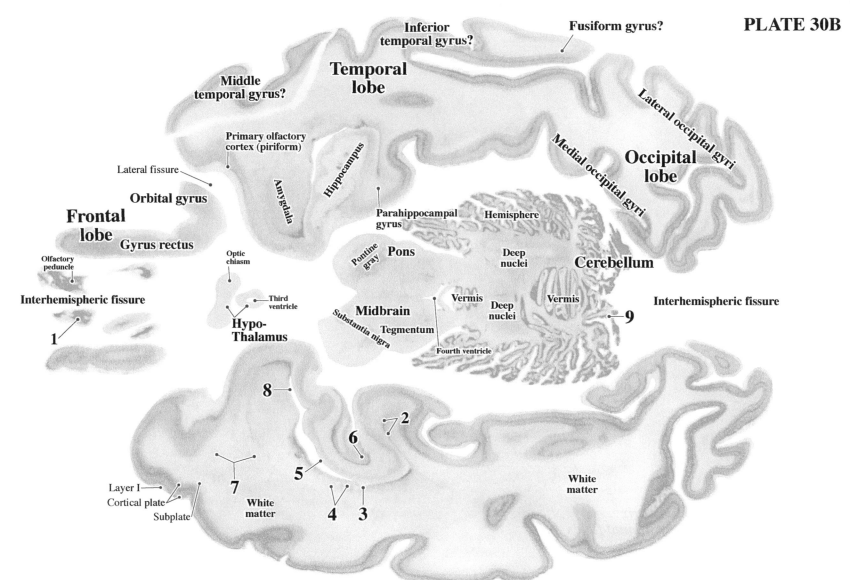

Fusiform gyrus?

Inferior
temporal gyrus?

Middle
temporal gyrus?

**Temporal
lobe**

Lateral occipital gyri

Primary olfactory
cortex (piriform)

Hippocampus

Medial occipital gyri

**Occipital
lobe**

Lateral fissure

Orbital gyrus

Amygdala

Parahippocampal
gyrus

Hemisphere

**Frontal
lobe**

Gyrus rectus

Optic
chiasm

Pontine
gray

Pons

Deep
nuclei

Cerebellum

Olfactory
peduncle

Third
ventricle

Vermis

Deep
nuclei

Vermis

Interhemispheric fissure

**Hypo-
Thalamus**

Midbrain

Substantia nigra

Tegmentum

Fourth ventricle

Interhemispheric fissure

1

8

2

6

5

7

White
matter

4 3

White
matter

Layer I

Cortical plate

Subplate

9

PLATE 31A
CR 320 mm
GW 37
Y301-62
Horizontal
Section 1111

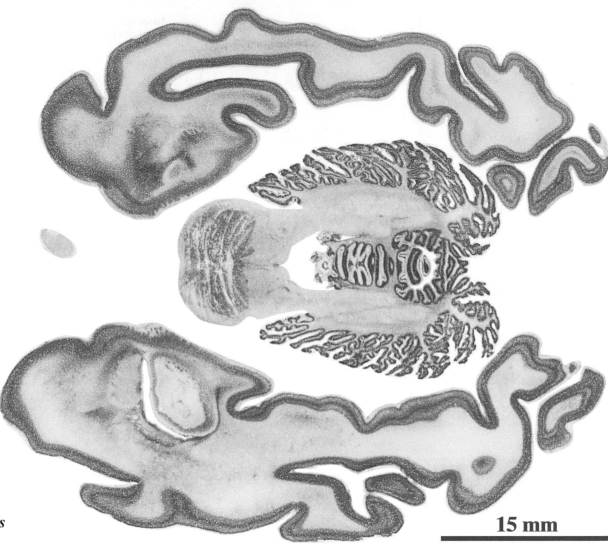

Remnants of the germinal matrix,
migratory streams, and transitional fields

1 *Parahippocampal NEP, SVZ, and STF*
2 *Temporal NEP and SVZ*
3 *Temporal STF*
4 *Alvear GEP*
5 *Lateral migratory stream (cortical)*
6 *Amygdaloid G/EP*
7 *External germinal layer (cerebellum)*

GEP - Glioepithelium
G/EP - Glioepithelium/ependyma
NEP - Neuroepithelium
STF - Stratified transitional field
SVZ - Subventricular zone

15 mm

See detail of brain core and
cerebellum in Plates 48A and B.

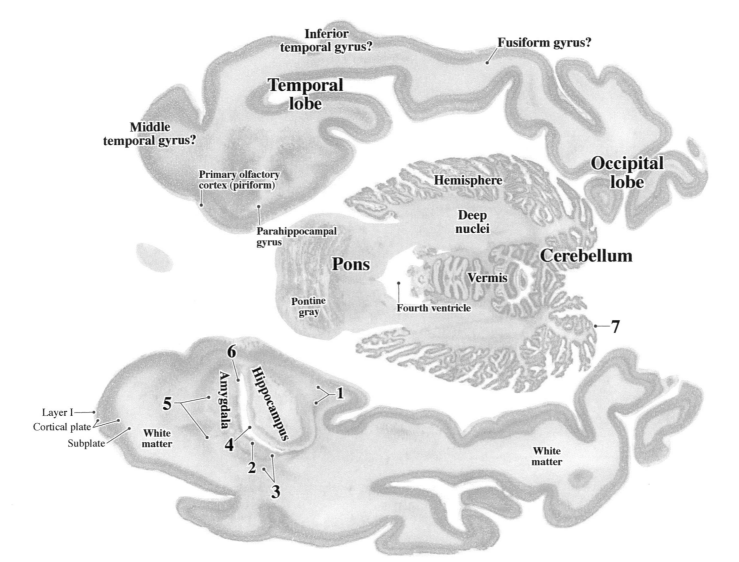

Inferior
temporal gyrus?

Fusiform gyrus?

**Temporal
lobe**

**Occipital
lobe**

Middle
temporal gyrus?

Hemisphere

Primary olfactory
cortex (piriform)

Deep
nuclei

Cerebellum

Parahippocampal
gyrus

Pons

Vermis

Pontine
gray

Fourth ventricle

7

6

Amygdala

Hippocampus

1

5

Layer I

Cortical plate

Subplate

White
matter

4

2

White
matter

3

PLATE 32A
CR 320 mm
GW 37
Y301-62
Horizontal
Section 1161

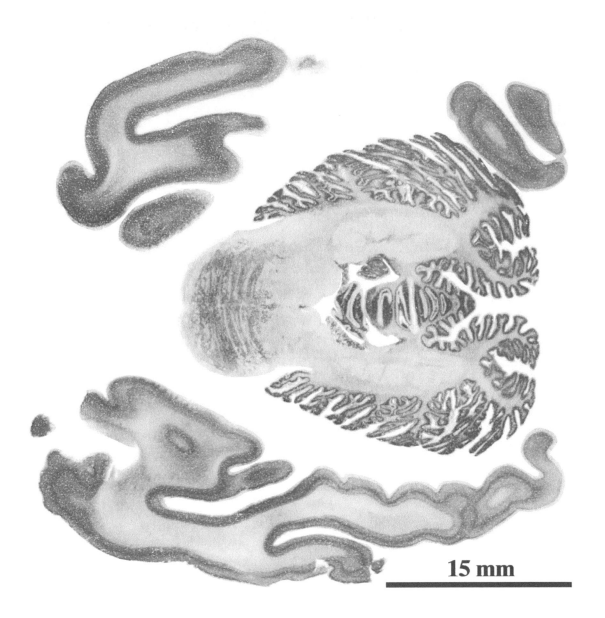

15 mm

See detail of brain core and
cerebellum in Plates 49A and B.

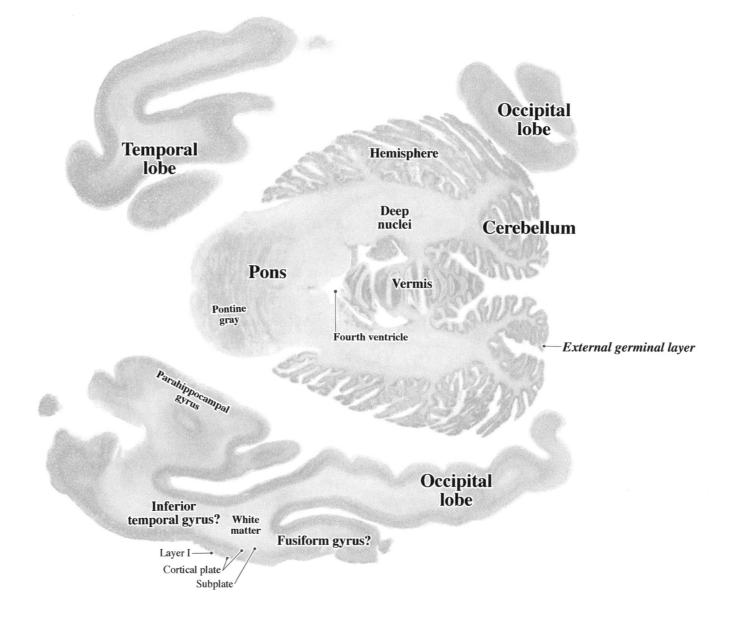

Temporal lobe

Occipital lobe

Hemisphere

Deep nuclei

Cerebellum

Pons

Vermis

Pontine gray

Fourth ventricle

External germinal layer

Parahippocampal gyrus

Occipital lobe

Inferior temporal gyrus?

White matter

Fusiform gyrus?

Layer I

Cortical plate

Subplate

PLATE 33A
CR 320 mm
GW 37
Y301-62
Horizontal
Section 1251

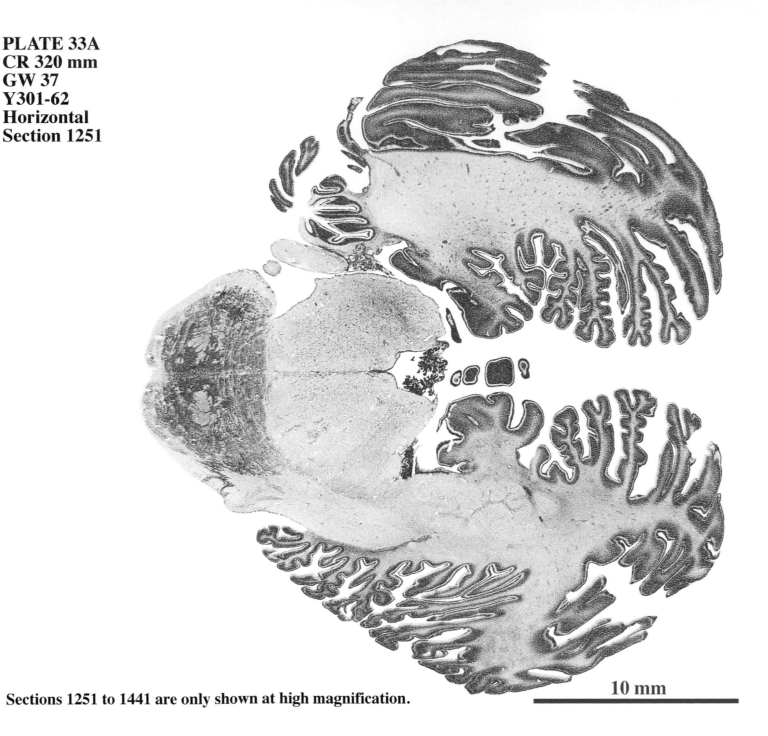

10 mm

Sections 1251 to 1441 are only shown at high magnification.

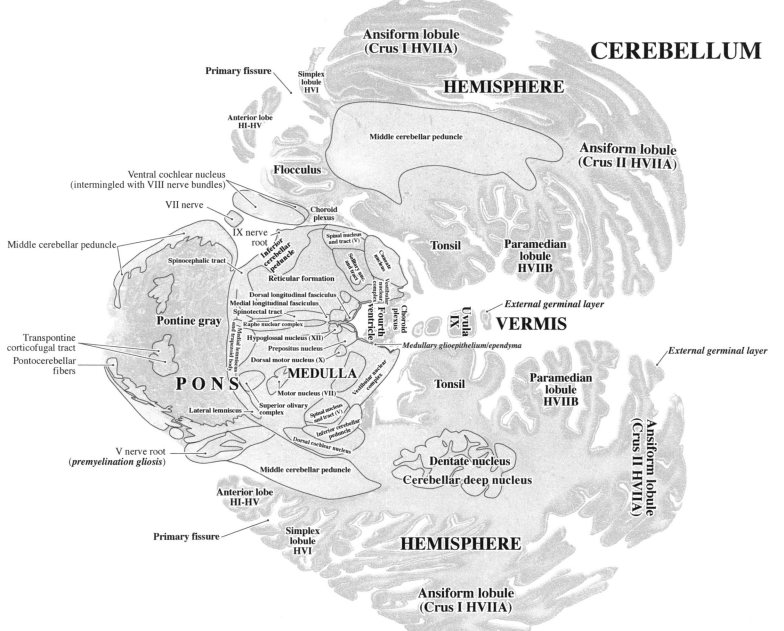

CEREBELLUM

Ansiform lobule
(Crus I HVIIA)

Primary fissure

Simplex lobule HVI

HEMISPHERE

Anterior lobe
HI-HV

Middle cerebellar peduncle

Ansiform lobule
(Crus II HVIIA)

Flocculus

Ventral cochlear nucleus
(intermingled with VIII nerve bundles)

Choroid plexus

VII nerve

IX nerve root

Spinal nucleus and tract (V)

Tonsil

Paramedian lobule HVIIB

Inferior cerebellar peduncle

Middle cerebellar peduncle

Spinocephalic tract

Cuneate nucleus

Solitary nuc. and tract

Reticular formation

Dorsal longitudinal fasciculus

Medial longitudinal fasciculus

Spinotectal tract

Raphe nuclear complex

Vestibular nuclear complex

Choroid plexus

External germinal layer

Uvula IX

VERMIS

Pontine gray

Medial lemniscus and trapezoid body

Hypoglossal nucleus (XII)

Prepositus nucleus

Dorsal motor nucleus (X)

Fourth Ventricle

Medullary glioepithelium/ependyma

External germinal layer

Transpontine corticofugal tract

Pontocerebellar fibers

PONS

MEDULLA

Motor nucleus (VII)

Vestibular nuclear complex

Tonsil

Paramedian lobule HVIIB

Lateral lemniscus

Superior olivary complex

Spinal nucleus and tract (V)

Ansiform lobule
(Crus II HVIIA)

Inferior cerebellar peduncle

Dorsal cochlear nucleus

V nerve root
(*premyelination gliosis*)

Middle cerebellar peduncle

Dentate nucleus

Cerebellar deep nucleus

Anterior lobe
HI-HV

Primary fissure

Simplex lobule HVI

HEMISPHERE

Ansiform lobule
(Crus I HVIIA)

Germinal and transitional structures in *italics*

PLATE 34A
CR 320 mm
GW 37
Y301-62
Horizontal
Section 1311

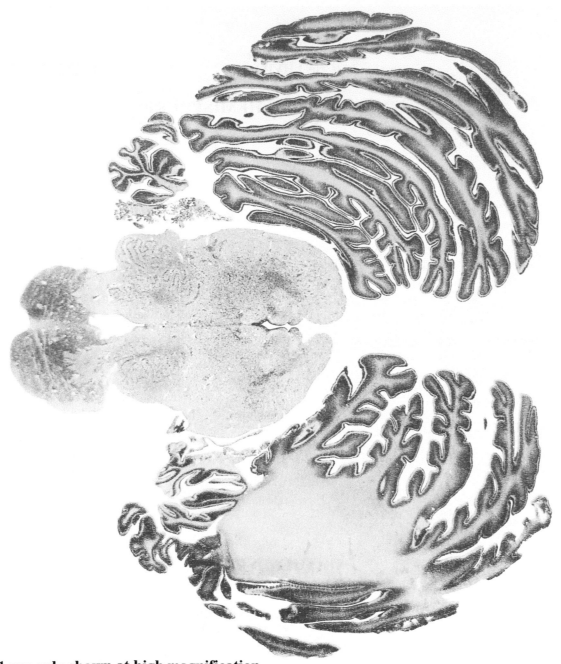

Sections 1251 to 1441 are only shown at high magnification.

CEREBELLUM

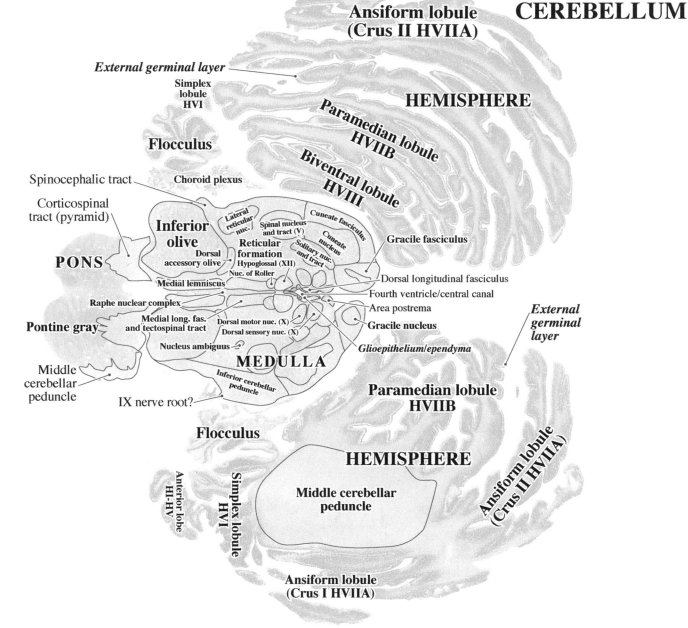

Ansiform lobule
(Crus II HVIIA)

External germinal layer

Simplex
lobule
HVI

HEMISPHERE

Paramedian lobule
HVIIB

Flocculus

Biventral lobule
HVIII

Spinocephalic tract

Choroid plexus

Cuneate fasciculus

Corticospinal
tract (pyramid)

Lateral
reticular
nuc.

Spinal nucleus
and tract (V)

Cuneate
nucleus

Gracile fasciculus

**Inferior
olive**

Reticular
formation

Solitary nuc.
and tract

Dorsal
accessory olive

PONS

Hypoglossal (XII)

Nuc. of Roller

Dorsal longitudinal fasciculus

Medial lemniscus

Fourth ventricle/central canal

Raphe nuclear complex

Area postrema

*External
germinal
layer*

Pontine gray

Medial long. fas.
and tectospinal tract

Dorsal motor nuc. (X)

Dorsal sensory nuc. (X)

Gracile nucleus

Nucleus ambiguus

Glioepithelium/ependyma

Middle
cerebellar
peduncle

MEDULLA

Inferior
cerebellar
peduncle

Paramedian lobule
HVIIB

IX nerve root?

Flocculus

HEMISPHERE

Anterior lobe
HI-HV

Simplex lobule
HVI

Middle cerebellar
peduncle

Ansiform lobule
(Crus II HVIIA)

Ansiform lobule
(Crus I HVIIA)

Germinal and transitional structures in *italics*

PLATE 35A
CR 320 mm
GW 37
Y301-62
Horizontal
Section 1331

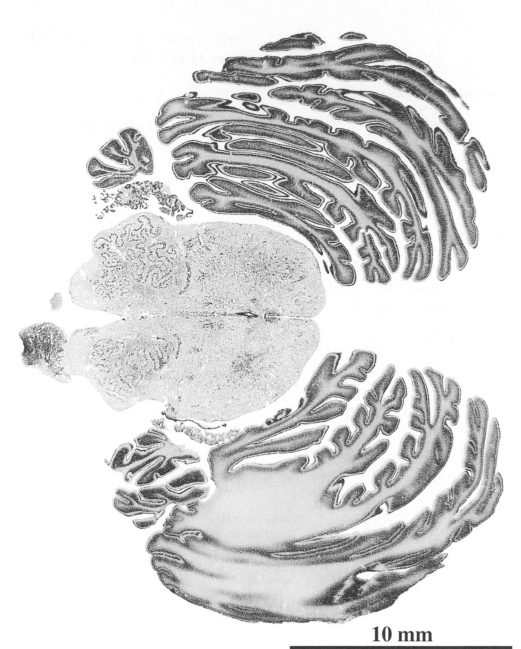

10 mm

Sections 1251 to 1441 are only shown at high magnification.

**Ansiform lobule
(Crus II HVIIA)**

CEREBELLUM

External germinal layer

**Paramedian lobule
HVIIB**

Flocculus

HEMISPHERE

Spinocerebellar tracts — Choroid plexus

*Biventral lobule
HVIII*

Spinocephalic tract

Lateral
reticular nuc.

Spinal nucleus
and tract (V)

Cuneate fasciculus

**Inferior
olive**

Nuc.
ambiguus

Cuneate
nucleus

Gracile fasciculus

Dorsal
accessory olive

Solitary nuc.
and tract

Corticospinal
tract (pyramid)

Pyramid

Nuc. of Roller

**Reticular
formation**

Central
canal

Medial lemniscus

Raphe nuc. complex

PONS

Medial long. fas.
and tectospinal tr.

Gracile
nucleus

Commissural nucleus (X)

Hypoglossal (XII)

Middle
cerebellar
peduncle

Dorsal motor nuc. (X)

Dorsal sensory nuc. (X)

Glioepithelium/ependyma

MEDULLA

X nerve root?

Inferior cerebellar
peduncle

**Paramedian lobule
HVIIB**

External germinal layer

Flocculus

**Middle cerebellar
peduncle**

Simplex
lobule
HVI

HEMISPHERE

**Ansiform lobule
(Crus II HVIIA)**

Germinal and transitional structures in *italics*

PLATE 36A
CR 320 mm
GW 37
Y301-62
Horizontal
Section 1381

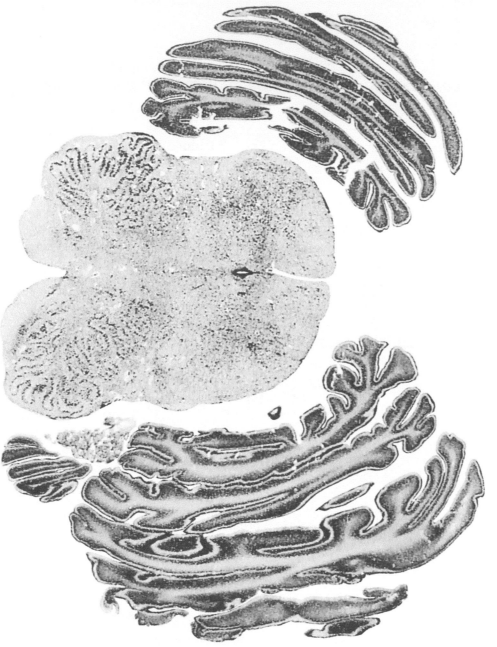

10 mm

Sections 1251 to 1441 are only shown at high magnification.

Paramedian lobule

CEREBELLUM

HEMISPHERE

Biventral lobule

MEDULLA

Medial longitudinal
fasciculus
and
tectospinal tract

External cuneate
nucleus

Cuneate nucleus

Cuneate fasciculus

Solitary nuc.
and tract

Pyramid

Nuc.
ambiguus

Dorsal motor
nuc. (X)

Central canal

Corticospinal
tract

**Medial lemniscus
(decussation)**

Raphe nuc.

Glioepithelium/ependyma

Medial accessory olive

Hypoglossal
nuc. (XII)

Gracile nucleus

Gracile fasciculus

Dorsal
accessory olive

**Reticular
formation**

Cuneate
nucleus

**Inferior
olive**

Lateral
reticular nuc.

Spinal
nucleus (V)

Cuneate fasciculus

External germinal layer

Spinal
tract (V)

Spinocerebellar tracts

Choroid plexus

Flocculus

Paramedian lobule

HEMISPHERE

**Ansiform lobule
(Crus I HVIIA)**

Germinal and transitional structures in *italics*

PLATE 37A
CR 320 mm
GW 37
Y301-62
Horizontal
Section 1441

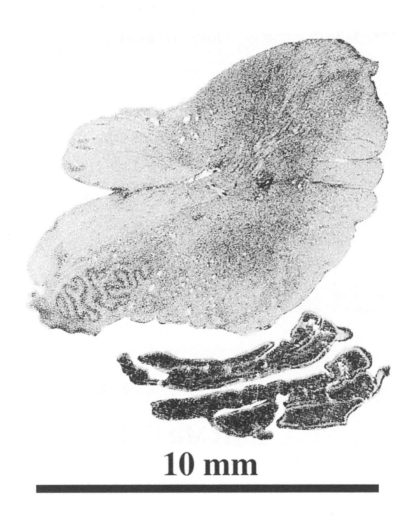

10 mm

Sections 1251 to 1441 are only shown at high magnification.

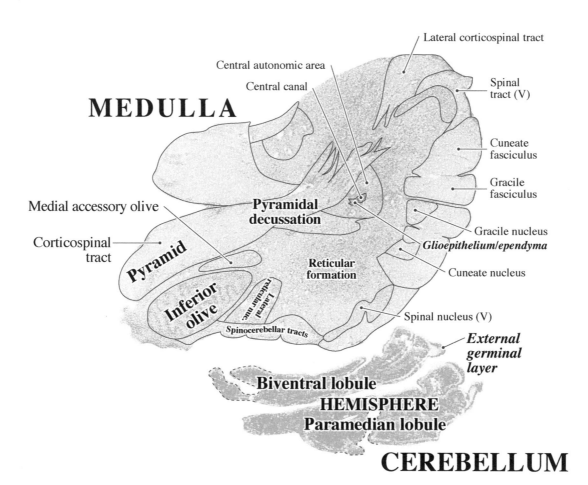

Lateral corticospinal tract

Central autonomic area

Central canal

Spinal tract (V)

MEDULLA

Cuneate fasciculus

Gracile fasciculus

Pyramidal decussation

Medial accessory olive

Gracile nucleus

Glioepithelium/ependyma

Corticospinal tract

Reticular formation

Cuneate nucleus

Pyramid

Lateral reticular nuc.

Inferior olive

Spinocerebellar tracts

Spinal nucleus (V)

External germinal layer

Biventral lobule

HEMISPHERE

Paramedian lobule

CEREBELLUM

Germinal and transitional structures in *italics*

PLATE 38A
CR 320 mm
GW 37
Y301-62
Horizontal
Section 531

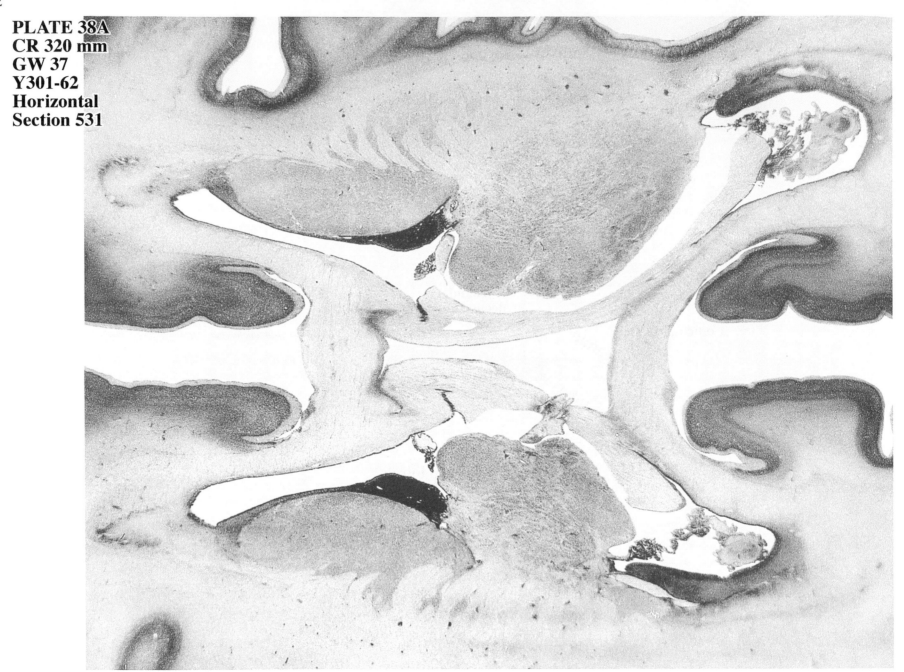

See the entire Section 531 in Plate 21.

10 mm

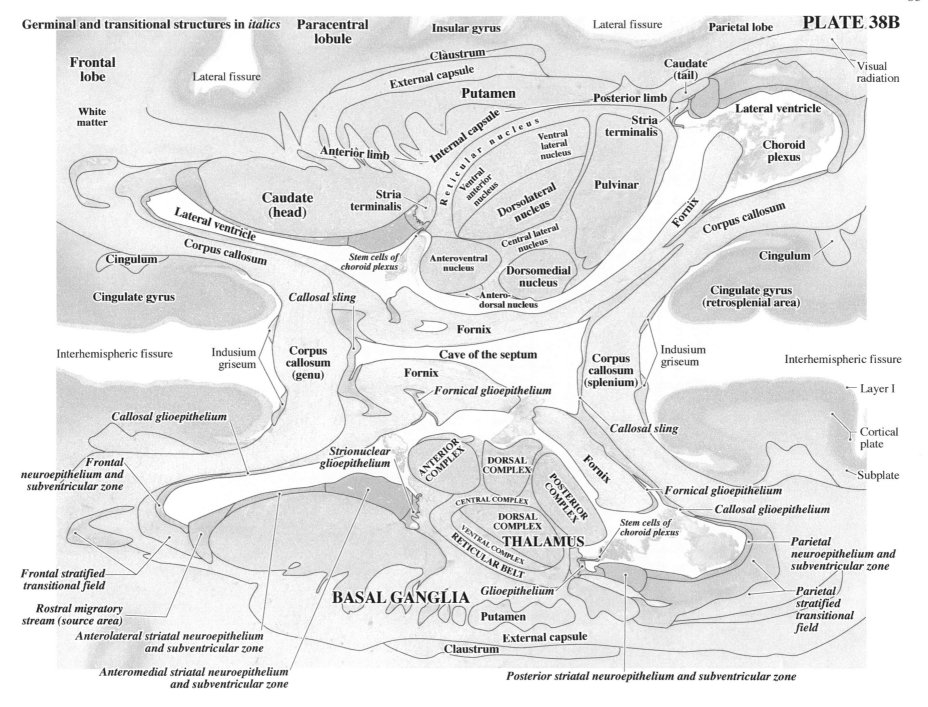

Germinal and transitional structures in *italics*

PLATE 38B

Frontal lobe

Paracentral lobule

Insular gyrus

Lateral fissure

Parietal lobe

Lateral fissure

Claustrum

External capsule

Caudate (tail)

Visual radiation

White matter

Putamen

Posterior limb

Stria terminalis

Lateral ventricle

Internal capsule

R e t i c u l a r n u c l e u s

Ventral lateral nucleus

Choroid plexus

Anterior limb

Ventral anterior nucleus

Pulvinar

Caudate (head)

Stria terminalis

Dorsolateral nucleus

Lateral ventricle

Stem cells of choroid plexus

Central lateral nucleus

Corpus callosum

Corpus callosum

Anteroventral nucleus

Fornix

Cingulum

Cingulum

Cingulate gyrus

Callosal sling

Dorsomedial nucleus

Antero-dorsal nucleus

Cingulate gyrus (retrosplenial area)

Fornix

Interhemispheric fissure

Indusium griseum

Corpus callosum (genu)

Cave of the septum

Corpus callosum (splenium)

Indusium griseum

Interhemispheric fissure

Fornix

Fornical glioepithelium

Layer I

Callosal glioepithelium

Callosal sling

Cortical plate

Frontal neuroepithelium and subventricular zone

Strionuclear glioepithelium

ANTERIOR COMPLEX

DORSAL COMPLEX

POSTERIOR COMPLEX

Fornix

Fornical glioepithelium

Subplate

CENTRAL COMPLEX

Callosal glioepithelium

Frontal stratified transitional field

DORSAL COMPLEX

Stem cells of choroid plexus

Parietal neuroepithelium and subventricular zone

VENTRAL COMPLEX

THALAMUS

RETICULAR BELT

Rostral migratory stream (source area)

BASAL GANGLIA

Glioepithelium

Parietal stratified transitional field

Anterolateral striatal neuroepithelium and subventricular zone

Putamen

External capsule

Anteromedial striatal neuroepithelium and subventricular zone

Claustrum

Posterior striatal neuroepithelium and subventricular zone

PLATE 39A
CR 320 mm
GW 37
Y301-62
Horizontal
Section 561

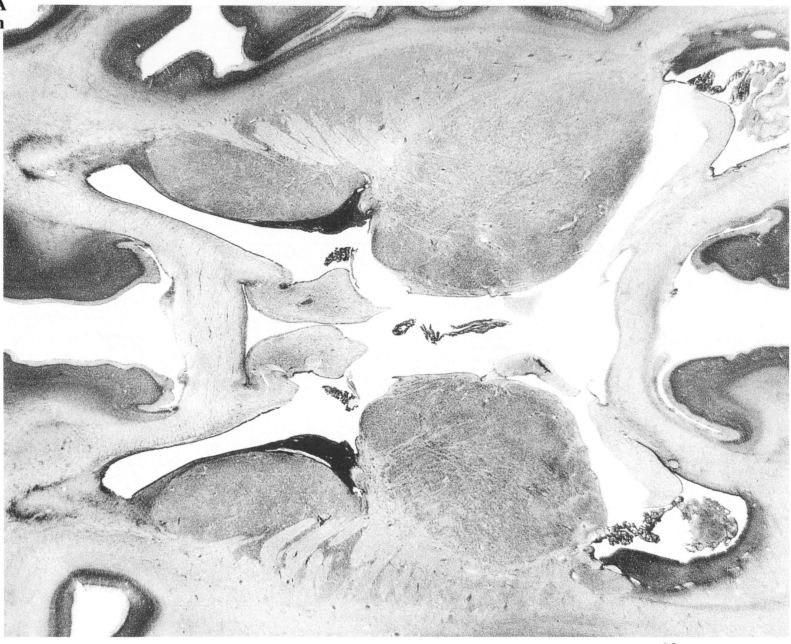

See the entire Section 561 in Plate 22.

10 mm

Germinal and transitional structures in *italics*

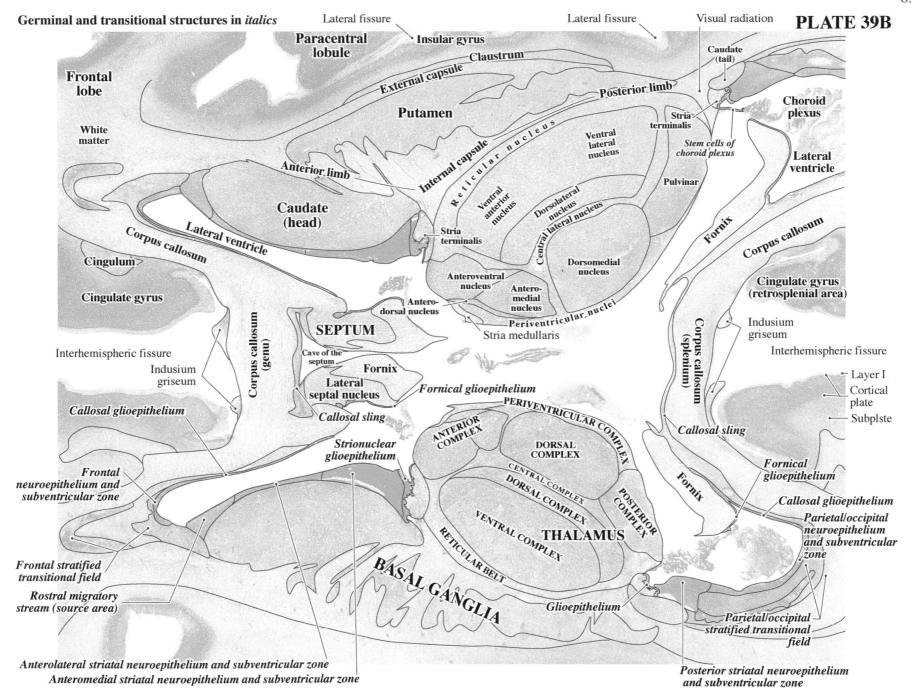

Lateral fissure

Insular gyrus

Paracentral lobule

Claustrum

External capsule

Frontal lobe

White matter

Putamen

Anterior limb

Internal capsule

Caudate (head)

R e t i c u l a r   n u c l e u s

Ventral anterior nucleus

Stria terminalis

Corpus callosum

Lateral ventricle

Cingulum

Cingulate gyrus

Interhemispheric fissure

Indusium griseum

Callosal glioepithelium

Corpus callosum (genu)

Cave of the septum

SEPTUM

Fornix

Lateral septal nucleus

Callosal sling

Strionuclear glioepithelium

Frontal neuroepithelium and subventricular zone

Frontal stratified transitional field

Rostral migratory stream (source area)

Lateral fissure

Visual radiation

Caudate (tail)

Posterior limb

Stria terminalis

Stem cells of choroid plexus

Ventral lateral nucleus

Pulvinar

Dorsolateral nucleus

Central lateral nucleus

Dorsomedial nucleus

Anteroventral nucleus

Antero-medial nucleus

Antero-dorsal nucleus

Periventricular nuclei

Stria medullaris

Fornical glioepithelium

PERIVENTRICULAR COMPLEX

ANTERIOR COMPLEX

DORSAL COMPLEX

CENTRAL COMPLEX

DORSAL COMPLEX

VENTRAL COMPLEX

POSTERIOR COMPLEX

THALAMUS

RETICULAR BELT

BASAL GANGLIA

Glioepithelium

Choroid plexus

Lateral ventricle

Fornix

Corpus callosum

Cingulate gyrus (retrosplenial area)

Indusium griseum

Interhemispheric fissure

Layer I

Cortical plate

Subplste

Corpus callosum (splenium)

Callosal sling

Fornix

Fornical glioepithelium

Callosal glioepithelium

Parietal/occipital neuroepithelium and subventricular zone

Parietal/occipital stratified transitional field

Posterior striatal neuroepithelium and subventricular zone

Anterolateral striatal neuroepithelium and subventricular zone

Anteromedial striatal neuroepithelium and subventricular zone

PLATE 40A
CR 320 mm
GW 37
Y301-62
Horizontal
Section 631

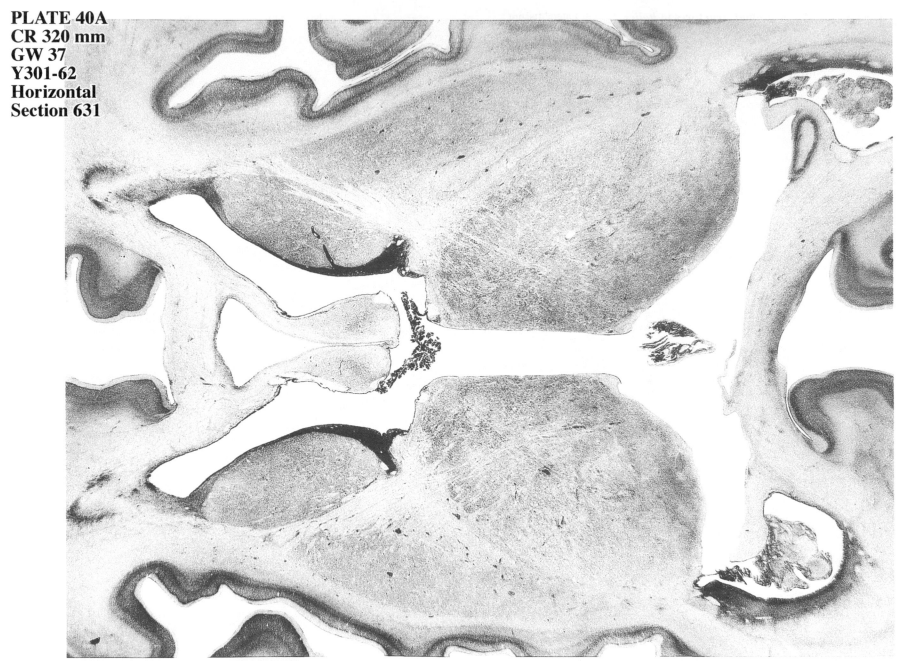

See the entire Section 631 in Plate 23.

10 mm

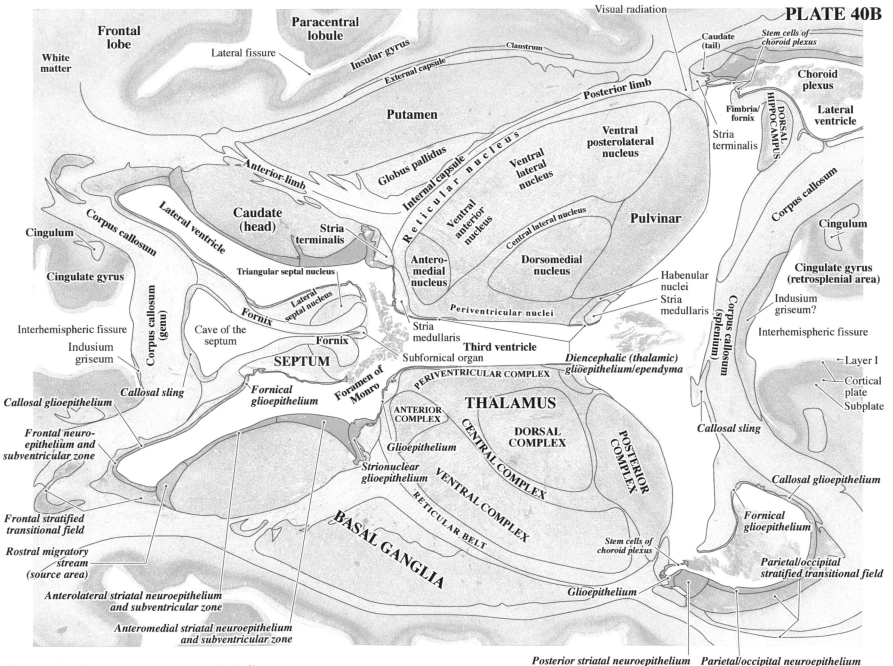

Germinal and transitional structures in *italics*

PLATE 41A
CR 320 mm
GW 37
Y301-62
Horizontal
Section 731

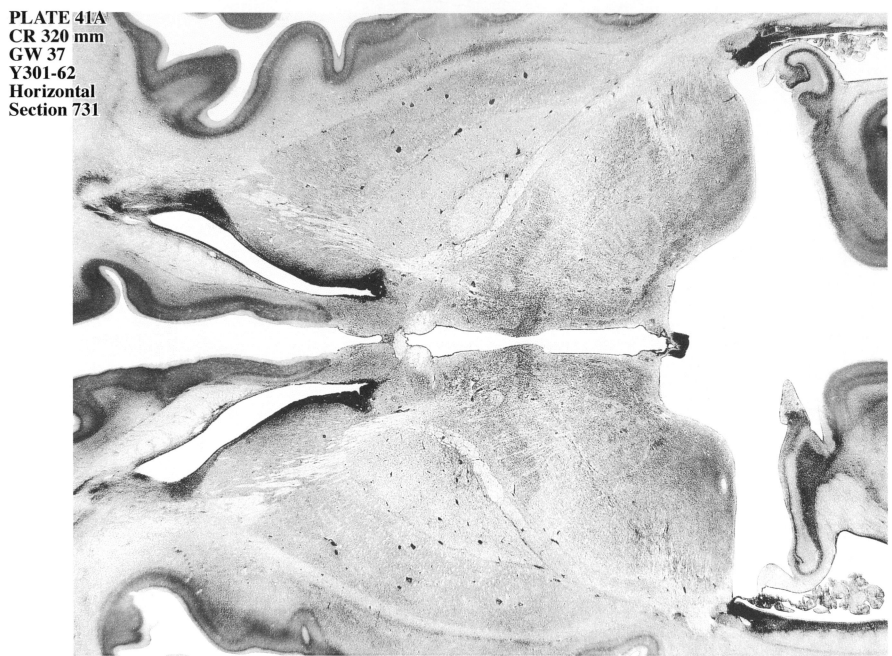

See the entire Section 731 in Plate 24.

10 mm

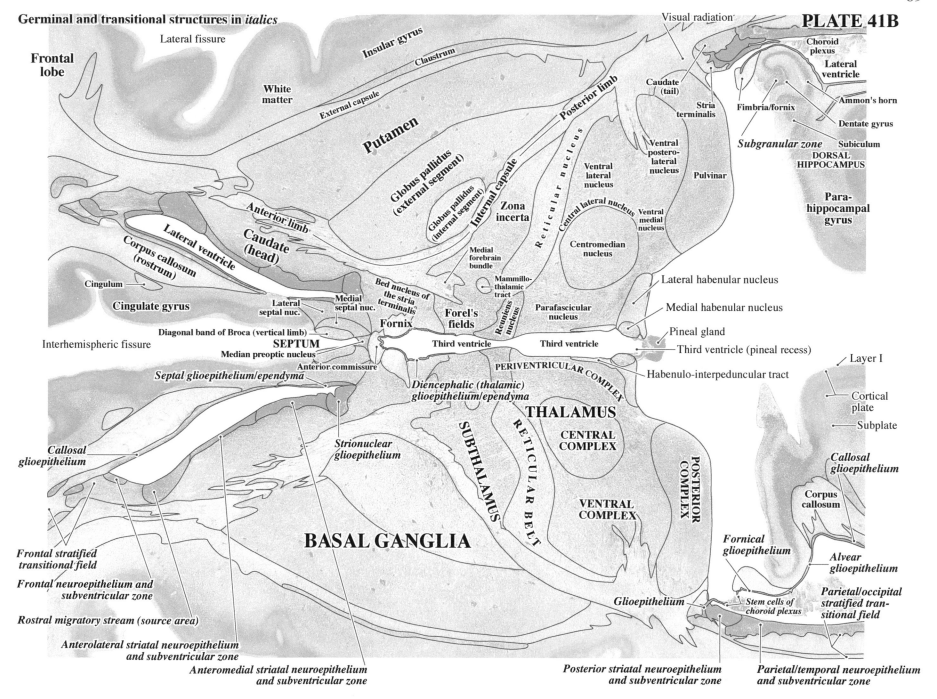

Germinal and transitional structures in *italics*

PLATE 41B

Lateral fissure

Frontal lobe

Insular gyrus

Visual radiation

Claustrum

Choroid plexus

Lateral ventricle

White matter

External capsule

Caudate (tail)

Stria terminalis

Fimbria/fornix

Ammon's horn

Putamen

Posterior limb

Ventral postero-lateral nucleus

Dentate gyrus

Subgranular zone

Subiculum

Globus pallidus (external segment)

Ventral lateral nucleus

Pulvinar

DORSAL HIPPOCAMPUS

Anterior limb

Globus pallidus (internal segment)

Internal capsule

Zona incerta

Reticular nucleus

Central lateral nucleus

Ventral medial nucleus

Para-hippocampal gyrus

Caudate (head)

Corpus callosum (rostrum)

Lateral ventricle

Medial forebrain bundle

Centromedian nucleus

Cingulum

Bed nucleus of the stria terminalis

Mammillo-thalamic tract

Lateral habenular nucleus

Cingulate gyrus

Lateral septal nuc.

Medial septal nuc.

Forel's fields

Parafascicular nucleus

Reuniens nucleus

Medial habenular nucleus

Diagonal band of Broca (vertical limb)

Fornix

Pineal gland

Interhemispheric fissure

SEPTUM

Median preoptic nucleus

Third ventricle

Third ventricle

Third ventricle (pineal recess)

Anterior commissure

PERIVENTRICULAR COMPLEX

Habenulo-interpeduncular tract

Layer I

Septal glioepithelium/ependyma

Diencephalic (thalamic) glioepithelium/ependyma

THALAMUS

Cortical plate

Callosal glioepithelium

Strionuclear glioepithelium

SUBTHALAMUS

RETICULAR BELT

CENTRAL COMPLEX

Subplate

POSTERIOR COMPLEX

Callosal glioepithelium

Corpus callosum

VENTRAL COMPLEX

Fornical glioepithelium

Alvear glioepithelium

Frontal stratified transitional field

BASAL GANGLIA

Parietal/occipital stratified transitional field

Frontal neuroepithelium and subventricular zone

Glioepithelium

Stem cells of choroid plexus

Rostral migratory stream (source area)

Anterolateral striatal neuroepithelium and subventricular zone

Anteromedial striatal neuroepithelium and subventricular zone

Posterior striatal neuroepithelium and subventricular zone

Parietal/temporal neuroepithelium and subventricular zone

PLATE 42A
CR 320 mm
GW 37
Y301-62
Horizontal
Section 761

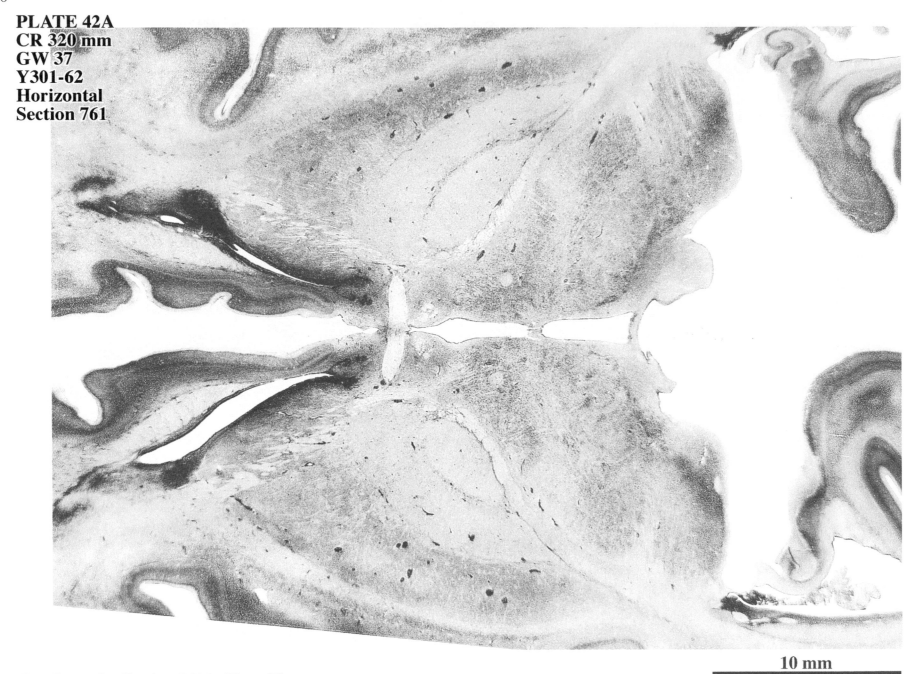

10 mm

See the entire Section 761 in Plate 25.

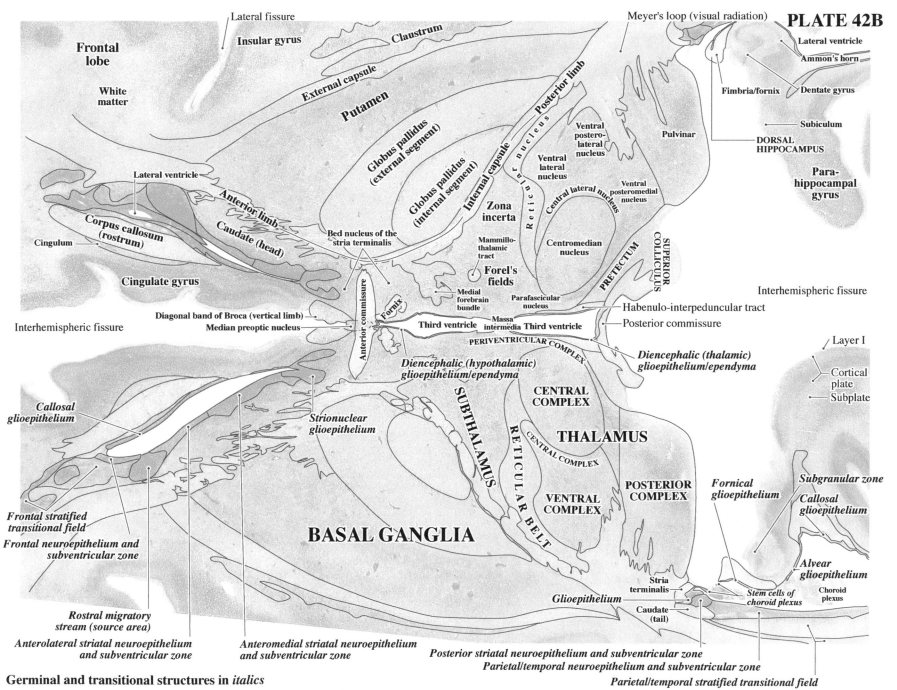

91

PLATE 42B

Lateral fissure

Insular gyrus

Claustrum

Meyer's loop (visual radiation)

Frontal lobe

White matter

External capsule

Putamen

Lateral ventricle

Ammon's horn

Dentate gyrus

Fimbria/fornix

Subiculum

DORSAL HIPPOCAMPUS

Globus pallidus (external segment)

Posterior limb

Pulvinar

Ventral postero-lateral nucleus

Para-hippocampal gyrus

Lateral ventricle

Corpus callosum (rostrum)

Anterior limb

Caudate (head)

Globus pallidus (internal segment)

Internal capsule

Reticular nucleus

Ventral lateral nucleus

Central lateral nucleus

Ventral posteromedial nucleus

Cingulum

Zona incerta

Bed nucleus of the stria terminalis

Cingulate gyrus

Forel's fields

Mammillo-thalamic tract

Centromedian nucleus

SUPERIOR COLLICULUS

PRETECTUM

Interhemispheric fissure

Medial forebrain bundle

Parafascicular nucleus

Anterior commissure

Fornix

Diagonal band of Broca (vertical limb)

Median preoptic nucleus

Habenulo-interpeduncular tract

Interhemispheric fissure

Third ventricle

Massa intermedia

Third ventricle

Posterior commissure

PERIVENTRICULAR COMPLEX

Diencephalic (thalamic) glioepithelium/ependyma

Diencephalic (hypothalamic) glioepithelium/ependyma

Layer I

Cortical plate

Subplate

Callosal glioepithelium

Strionuclear glioepithelium

SUBTHALAMUS

CENTRAL COMPLEX

THALAMUS

Fornical glioepithelium

Subgranular zone

Callosal glioepithelium

RETICULAR BELT

CENTRAL COMPLEX

Frontal stratified transitional field

Frontal neuroepithelium and subventricular zone

VENTRAL COMPLEX

POSTERIOR COMPLEX

Alvear glioepithelium

BASAL GANGLIA

Stria terminalis

Stem cells of choroid plexus

Choroid plexus

Rostral migratory stream (source area)

Glioepithelium

Caudate (tail)

Anterolateral striatal neuroepithelium and subventricular zone

Anteromedial striatal neuroepithelium and subventricular zone

Posterior striatal neuroepithelium and subventricular zone

Parietal/temporal neuroepithelium and subventricular zone

Germinal and transitional structures in *italics*

Parietal/temporal stratified transitional field

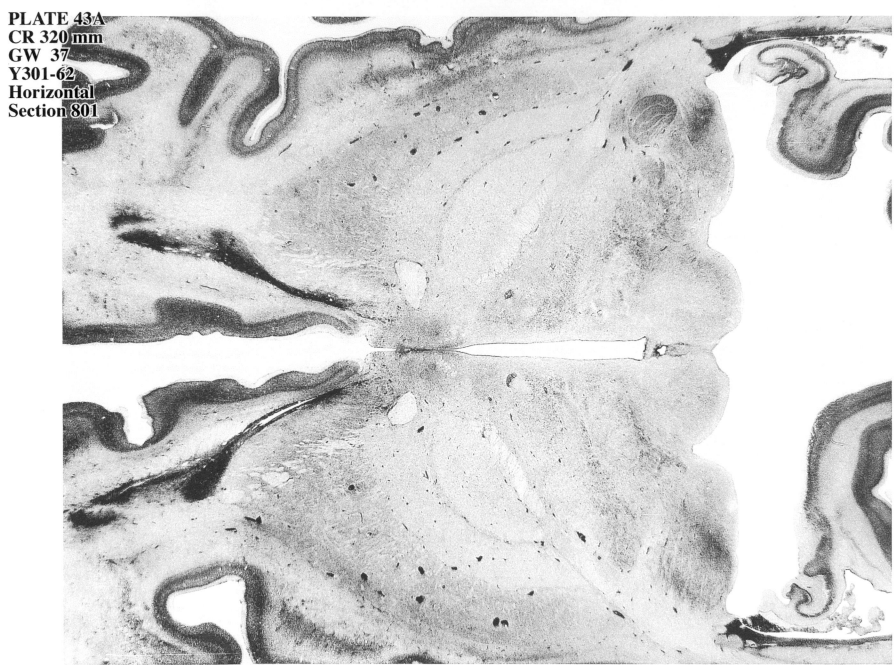

PLATE 43A
CR 320 mm
GW 37
Y301-62
Horizontal
Section 801

See the entire Section 801 in Plate 26.

10 mm

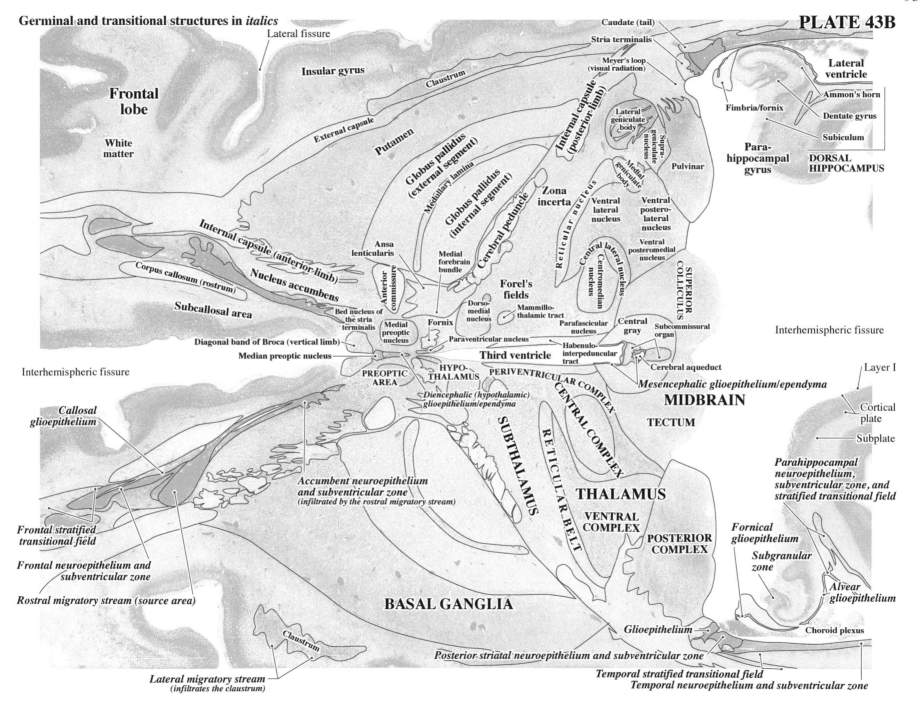

Germinal and transitional structures in *italics*

PLATE 43B

Lateral fissure

Caudate (tail)

Stria terminalis

Meyer's loop
(visual radiation)

Lateral ventricle

Insular gyrus

Claustrum

Ammon's horn

Frontal lobe

External capsule

Lateral geniculate body

Supra-geniculate nucleus

Fimbria/fornix

Dentate gyrus

Subiculum

White matter

Putamen

Globus pallidus (external segment)

Medial geniculate body

Pulvinar

Para-hippocampal gyrus

DORSAL HIPPOCAMPUS

Medullary lamina

Globus pallidus (internal segment)

Zona incerta

Internal capsule (posterior limb)

Reticular nucleus

Ventral lateral nucleus

Ventral postero-lateral nucleus

SUPERIOR COLLICULUS

Internal capsule (anterior limb)

Cerebral peduncle

Ventral posteromedial nucleus

Corpus callosum (rostrum)

Nucleus accumbens

Ansa lenticularis

Medial forebrain bundle

Central lateral nucleus

Centromedian nucleus

Subcallosal area

Anterior commissure

Forel's fields

Dorso-medial nucleus

Mammillo-thalamic tract

Parafascicular nucleus

Central gray

Subcommissural organ

Bed nucleus of the stria terminalis

Fornix

Paraventricular nucleus

Habenulo-interpeduncular tract

Cerebral aqueduct

Interhemispheric fissure

Diagonal band of Broca (vertical limb)

Medial preoptic nucleus

Median preoptic nucleus

Third ventricle

HYPO-THALAMUS

PERIVENTRICULAR COMPLEX

Mesencephalic glioepithelium/ependyma

Layer I

Interhemispheric fissure

PREOPTIC AREA

Diencephalic (hypothalamic) glioepithelium/ependyma

CENTRAL COMPLEX

MIDBRAIN

Cortical plate

Callosal glioepithelium

SUBTHALAMUS

RETICULAR BELT

TECTUM

Subplate

Parahippocampal neuroepithelium, subventricular zone, and stratified transitional field

Accumbent neuroepithelium and subventricular zone
(*infiltrated by the rostral migratory stream*)

THALAMUS

Frontal stratified transitional field

VENTRAL COMPLEX

Fornical glioepithelium

Frontal neuroepithelium and subventricular zone

POSTERIOR COMPLEX

Subgranular zone

Rostral migratory stream (source area)

Alvear glioepithelium

BASAL GANGLIA

Glioepithelium

Choroid plexus

Claustrum

Posterior-striatal neuroepithelium and subventricular zone

Lateral migratory stream
(*infiltrates the claustrum*)

Temporal stratified transitional field

Temporal neuroepithelium and subventricular zone

94

PLATE 44A
CR 320 mm
GW 37
Y301-62
Horizontal
Section 851

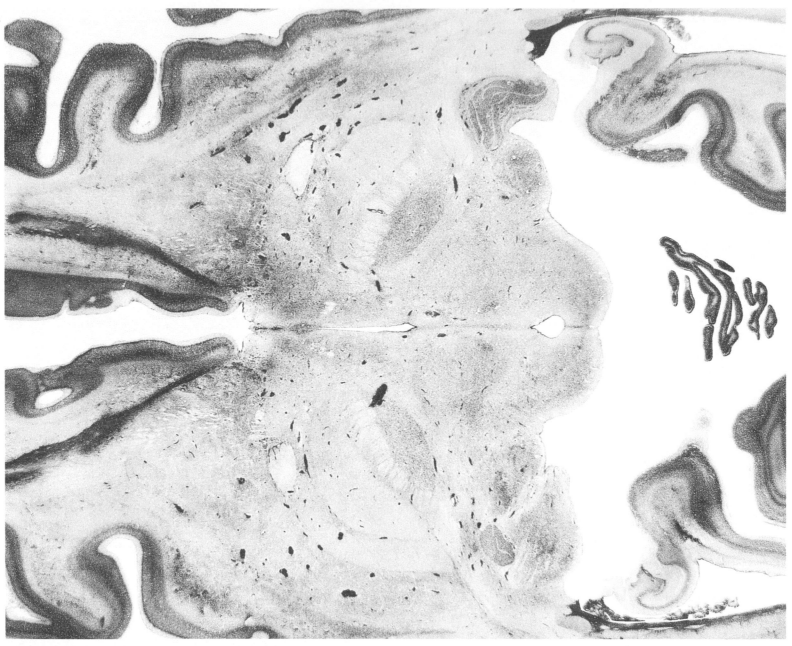

10 mm

See the entire Section 851 in Plate 27.

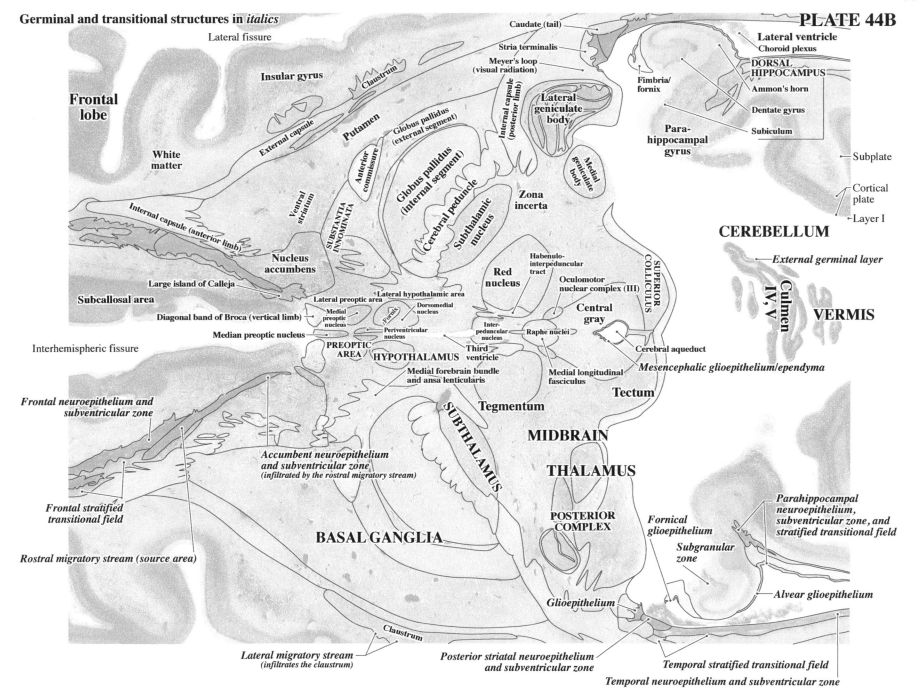

Germinal and transitional structures in *italics*

Lateral fissure

Caudate (tail)

Stria terminalis

Meyer's loop
(visual radiation)

Lateral ventricle
Choroid plexus

Insular gyrus

Claustrum

Internal capsule
(posterior limb)

**Lateral
geniculate
body**

Fimbria/
fornix

**DORSAL
HIPPOCAMPUS**

Ammon's horn

**Frontal
lobe**

External capsule

Putamen

Globus pallidus
(external segment)

Medial
geniculate
body

Para-
hippocampal
gyrus

Dentate gyrus

Subiculum

Subplate

**White
matter**

Anterior
commissure

Globus pallidus
(internal segment)

Cortical
plate

Internal capsule (anterior limb)

Ventral
striatum

SUBSTANTIA
INNOMINATA

Cerebral peduncle

Subthalamic
nucleus

Zona
incerta

Layer I

CEREBELLUM

External germinal layer

**Nucleus
accumbens**

Red
nucleus

Habenulo-
interpeduncular
tract

Oculomotor
nuclear complex (III)

SUPERIOR
COLLICULUS

Culmen
IV, V

Large island of Calleja

Lateral hypothalamic area

**Central
gray**

VERMIS

Subcallosal area

Lateral preoptic area

Dorsomedial
nucleus

Diagonal band of Broca (vertical limb)

Medial
preoptic
nucleus

Fornix

Periventricular
nucleus

Inter-
peduncular
nucleus

Raphe nuclei

Cerebral aqueduct

Median preoptic nucleus

PREOPTIC
AREA

HYPOTHALAMUS

Third
ventricle

Medial longitudinal
fasciculus

Mesencephalic glioepithelium/ependyma

Interhemispheric fissure

Medial forebrain bundle
and ansa lenticularis

Tegmentum

Tectum

*Frontal neuroepithelium and
subventricular zone*

SUBTHALAMUS

MIDBRAIN

*Accumbent neuroepithelium
and subventricular zone
(infiltrated by the rostral migratory stream)*

THALAMUS

*Frontal stratified
transitional field*

**POSTERIOR
COMPLEX**

*Fornical
glioepithelium*

*Parahippocampal
neuroepithelium,
subventricular zone, and
stratified transitional field*

Rostral migratory stream (source area)

*Subgranular
zone*

BASAL GANGLIA

Alvear glioepithelium

Glioepithelium

Claustrum

*Lateral migratory stream
(infiltrates the claustrum)*

*Posterior striatal neuroepithelium
and subventricular zone*

Temporal stratified transitional field

Temporal neuroepithelium and subventricular zone

PLATE 45A
CR 320 mm
GW 37
Y301-62
Horizontal
Section 921

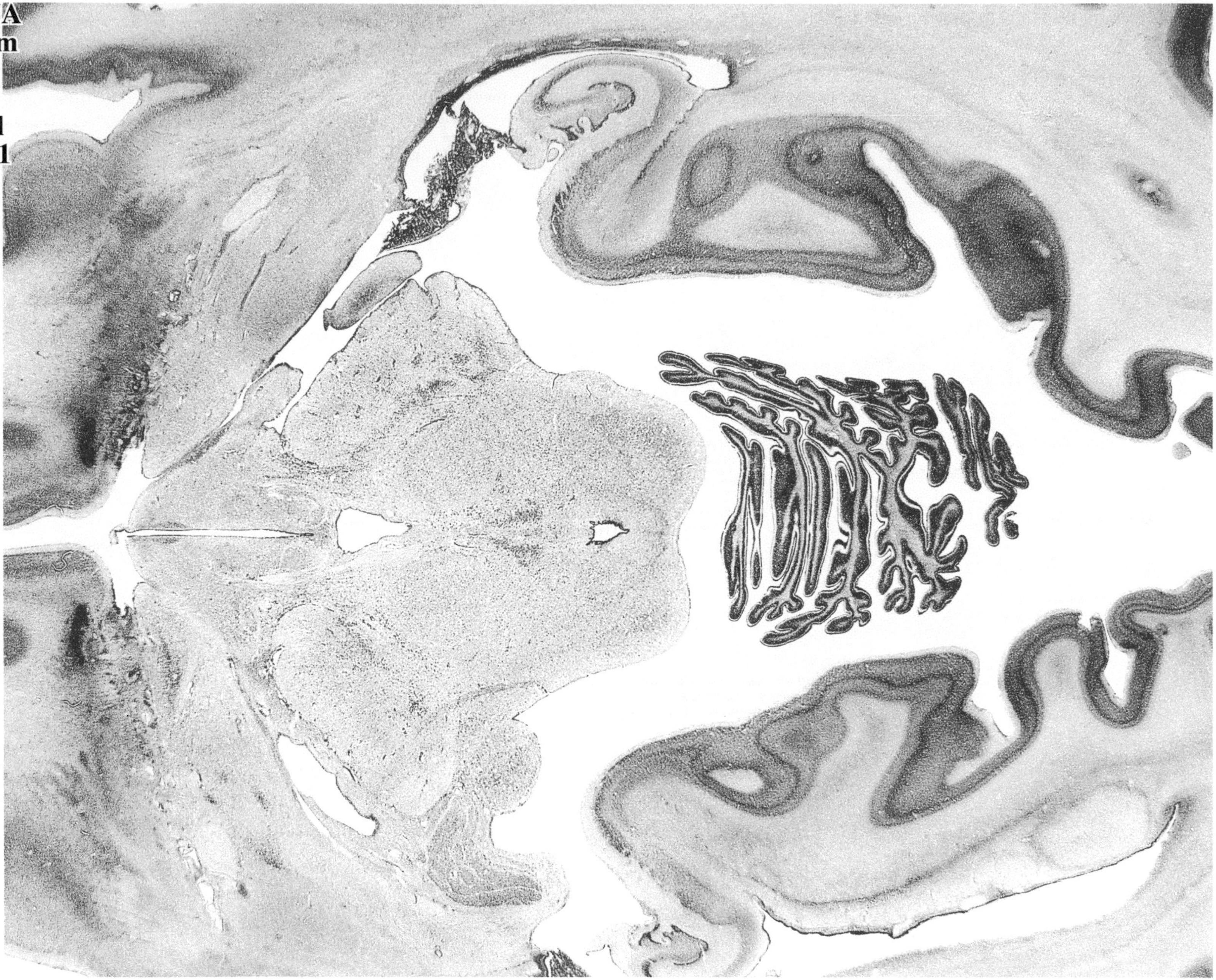

10 mm

See the entire Section 921 in Plate 28.

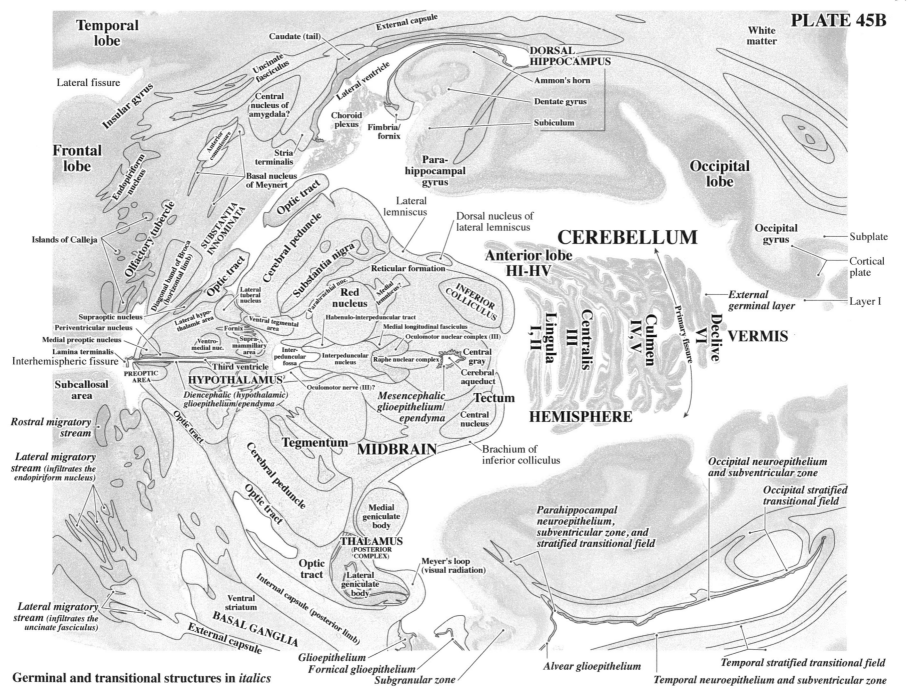

Temporal lobe

Lateral fissure

Insular gyrus

Frontal lobe

Endopiriform nucleus

Islands of Calleja

Olfactory tubercle

Rostral migratory stream

Lateral migratory stream (infiltrates the endopiriform nucleus)

Lateral migratory stream (infiltrates the uncinate fasciculus)

External capsule

Uncinate fasciculus

Caudate (tail)

External capsule

Central nucleus of amygdala?

Lateral ventricle

Choroid plexus

Anterior commissure

Stria terminalis

Basal nucleus of Meynert

SUBSTANTIA INNOMINATA

Diagonal band of Broca (horizontal limb)

Optic tract

Optic tract

Supraoptic nucleus
Periventricular nucleus
Medial preoptic nucleus
Lamina terminalis
Interhemispheric fissure

Subcallosal area

PREOPTIC AREA

Lateral tuberal nucleus

Ventral tegmental area

Lateral hypothalamic area

Fornix

Ventromedial nuc.

Supramammillary area

Supra-mammillary area

Third ventricle

HYPOTHALAMUS

Diencephalic (hypothalamic) glioepithelium/ependyma

Optic tract

Optic tract

Cerebral peduncle

Tegmentum

Cerebral peduncle

Optic tract

Ventral striatum

Medial geniculate body

THALAMUS (POSTERIOR COMPLEX)

Optic tract

Lateral geniculate body

Internal capsule (posterior limb)

BASAL GANGLIA

External capsule

Optic tract

Cerebral peduncle

Substantia nigra

Parabrachial nuc.

Red nucleus

Medial lemniscus?

Reticular formation

Habenulo-interpeduncular tract

Inter-peduncular fossa

Interpeduncular nucleus

Medial longitudinal fasciculus

Oculomotor nuclear complex (III)

Raphe nuclear complex

Oculomotor nerve (III)?

MIDBRAIN

Mesencephalic glioepithelium/ependyma

Central gray

Cerebral aqueduct

Tectum

Central nucleus

Brachium of inferior colliculus

Meyer's loop (visual radiation)

Lateral lemniscus

Dorsal nucleus of lateral lemniscus

INFERIOR COLLICULUS

Anterior lobe HI-HV

CEREBELLUM

Lingula I, II

Centralis III

Culmen IV, V

Primary fissure

Declive VI

VERMIS

HEMISPHERE

DORSAL HIPPOCAMPUS

Ammon's horn

Dentate gyrus

Subiculum

Para-hippocampal gyrus

Fimbria/fornix

Occipital lobe

White matter

Occipital gyrus

Subplate

Cortical plate

External germinal layer

Layer I

Occipital neuroepithelium and subventricular zone

Occipital stratified transitional field

Parahippocampal neuroepithelium, subventricular zone, and stratified transitional field

Temporal stratified transitional field

Alvear glioepithelium

Temporal neuroepithelium and subventricular zone

Glioepithelium
Fornical glioepithelium
Subgranular zone

Germinal and transitional structures in *italics*

PLATE 46A
CR 320 mm
GW 37
Y301-62
Horizontal
Section 981

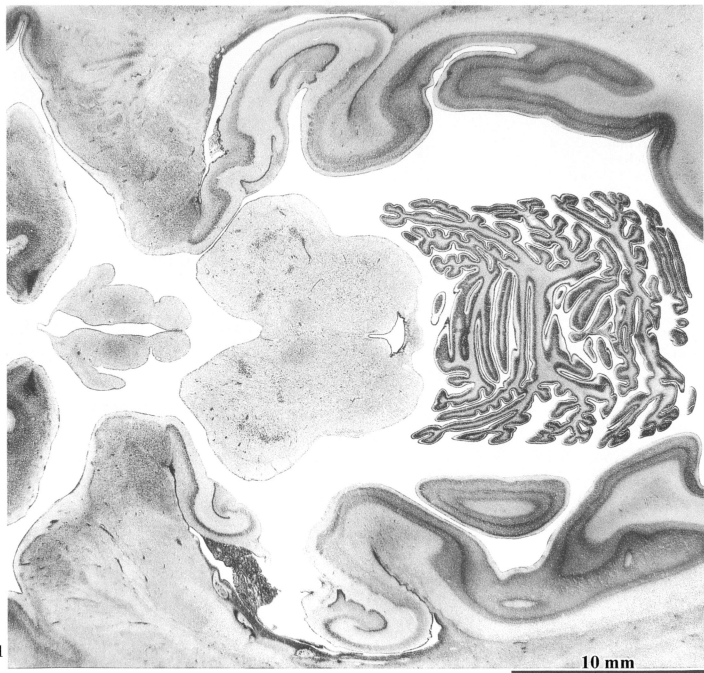

See the entire Section 981
in Plate 29.

10 mm

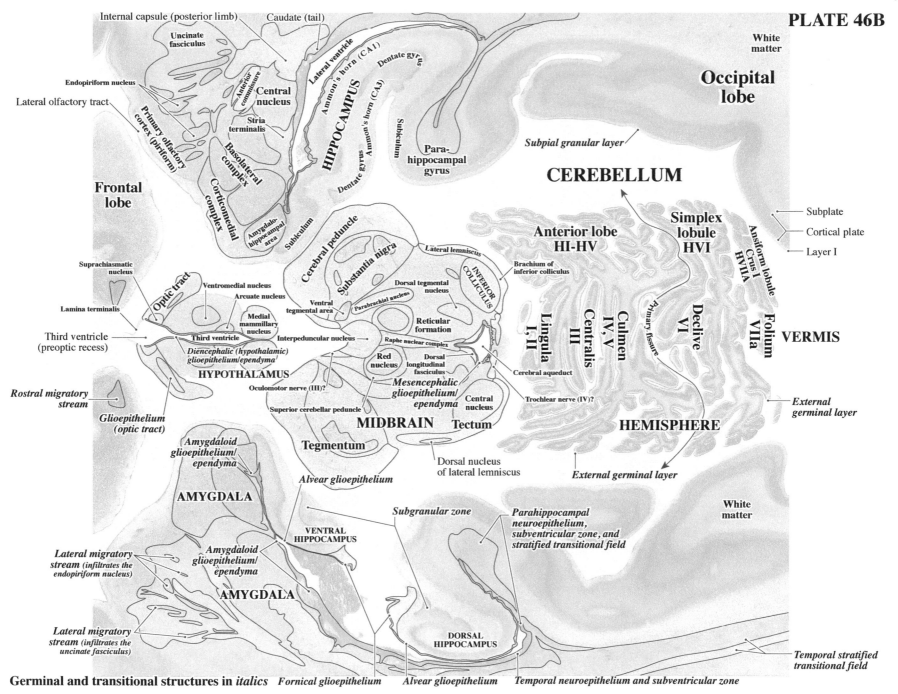

Internal capsule (posterior limb)

Caudate (tail)

White matter

Uncinate fasciculus

Occipital lobe

Endopiriform nucleus

Anterior commissure

Central nucleus

Lateral ventricle

Ammon's horn (CA1)

Dentate gyrus

Lateral olfactory tract

Primary olfactory cortex (piriform)

Stria terminalis

HIPPOCAMPUS

Ammon's horn (CA3)

Subiculum

Subpial granular layer

Frontal lobe

Basolateral complex

Dentate gyrus

Para-hippocampal gyrus

CEREBELLUM

Corticomedial complex

Amygdalo-hippocampal area

Subiculum

Anterior lobe HI–HV

Simplex lobule HVI

Ansiform lobule Crus I HVIIA

Subplate

Cortical plate

Suprachiasmatic nucleus

Cerebral peduncle

Substantia nigra

Lateral lemniscus

INFERIOR COLLICULUS

Brachium of inferior colliculus

Layer I

Ventromedial nucleus

Arcuate nucleus

Dorsal tegmental nucleus

Lingula I, II

Centralis III

Culmen IV, V

Primary fissure

Declive VI

Lamina terminalis

Optic tract

Medial mammillary nucleus

Ventral tegmental area

Parabrachial nucleus

Reticular formation

Folium VIIA

VERMIS

Third ventricle (preoptic recess)

Third ventricle

Interpeduncular nucleus

Raphe nuclear complex

Diencephalic (hypothalamic) glioepithelium/ependyma

HYPOTHALAMUS

Red nucleus

Dorsal longitudinal fasciculus

Cerebral aqueduct

Rostral migratory stream

Oculomotor nerve (III)?

Mesencephalic glioepithelium/ependyma

Central nucleus

Trochlear nerve (IV)?

External germinal layer

Glioepithelium (optic tract)

Superior cerebellar peduncle

MIDBRAIN

Tectum

HEMISPHERE

Amygdaloid glioepithelium/ependyma

Tegmentum

Dorsal nucleus of lateral lemniscus

External germinal layer

Alvear glioepithelium

AMYGDALA

Subgranular zone

Parahippocampal neuroepithelium, subventricular zone, and stratified transitional field

White matter

Lateral migratory stream (infiltrates the endopiriform nucleus)

Amygdaloid glioepithelium/ependyma

VENTRAL HIPPOCAMPUS

AMYGDALA

Lateral migratory stream (infiltrates the uncinate fasciculus)

DORSAL HIPPOCAMPUS

Temporal stratified transitional field

Germinal and transitional structures in *italics*   *Fornical glioepithelium*   *Alvear glioepithelium*   *Temporal neuroepithelium and subventricular zone*

**PLATE 47A
CR 320 mm
GW 37
Y301-62
Horizontal
Section 1051**

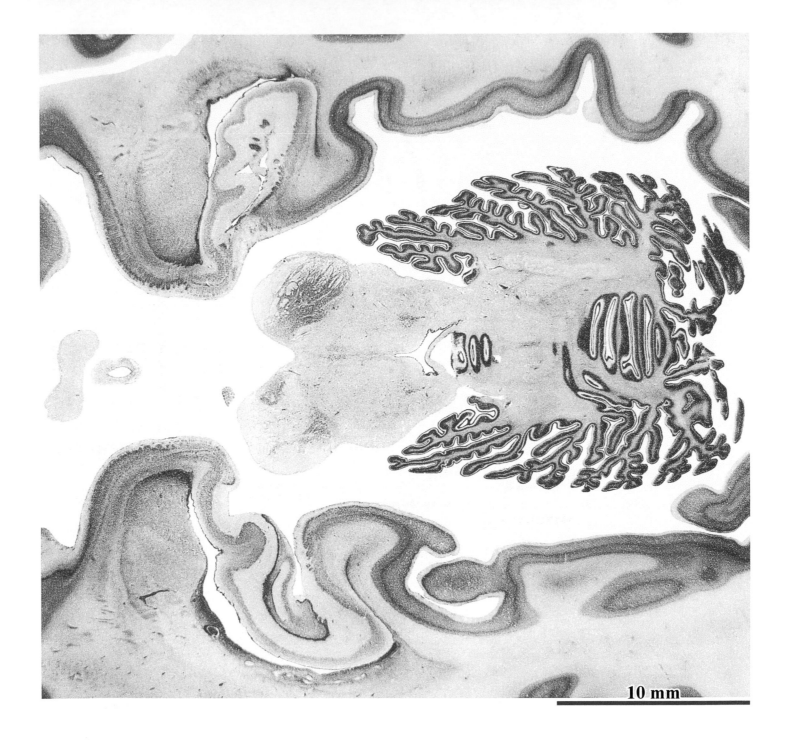

**See the entire
Section 1051
in Plate 30.**

10 mm

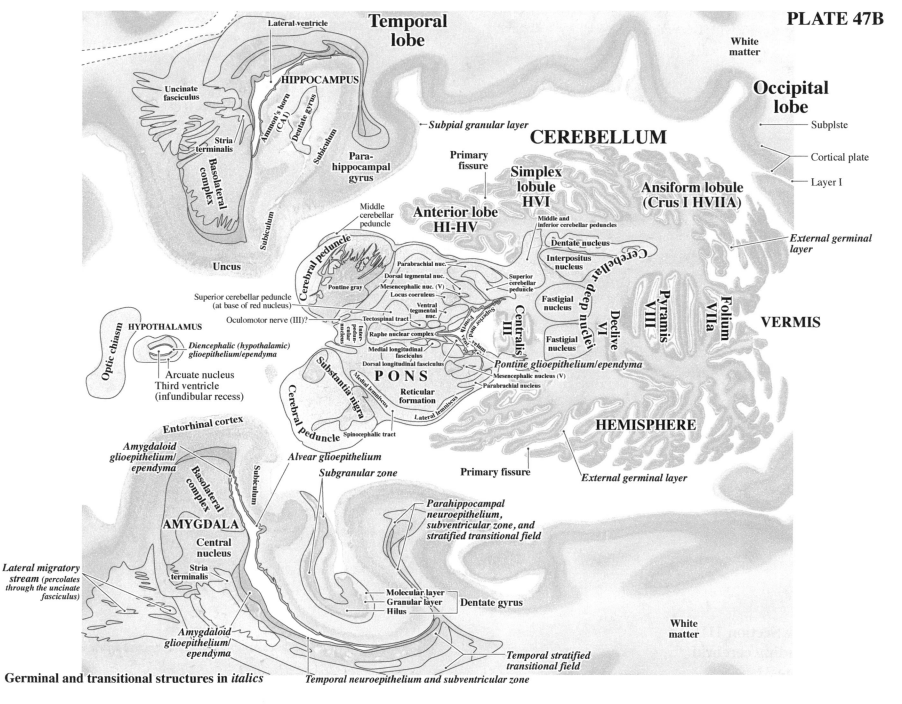

Germinal and transitional structures in *italics*

PLATE 48A
CR 320 mm
GW 37
Y301-62
Horizontal
Section 1111

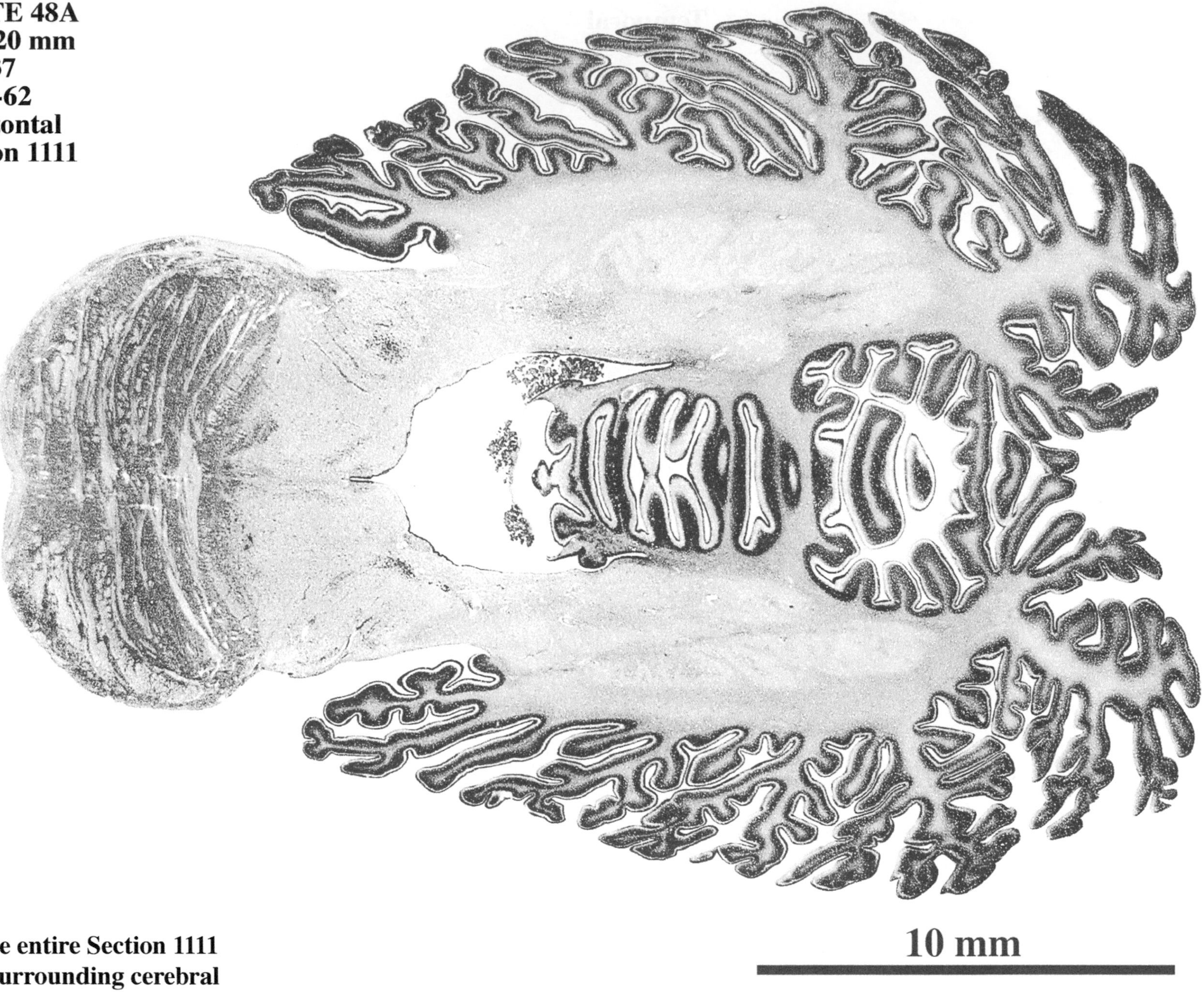

10 mm

See the entire Section 1111
with surrounding cerebral
cortex in Plate 31.

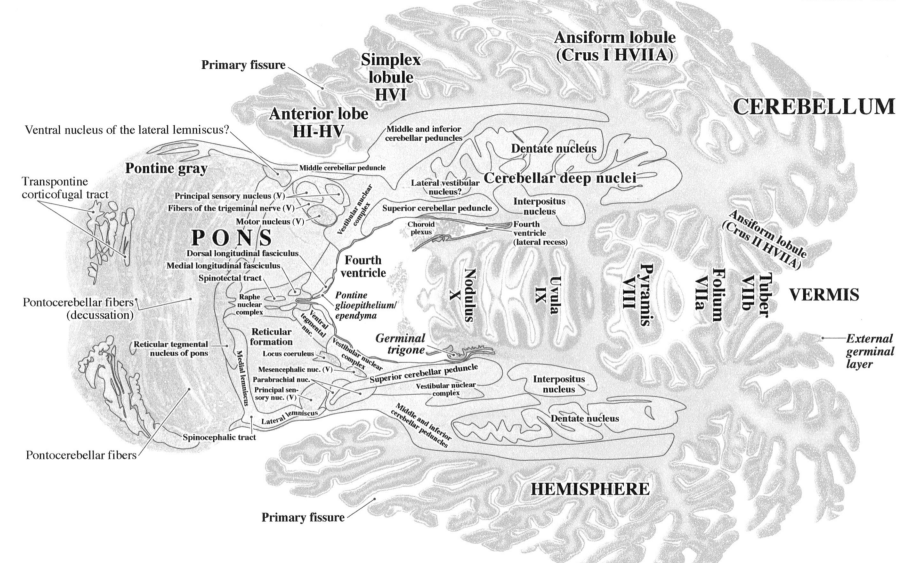

Germinal and transitional structures in *italics*

PLATE 49A
CR 320 mm
GW 37
Y301-62
Horizontal
Section 1161

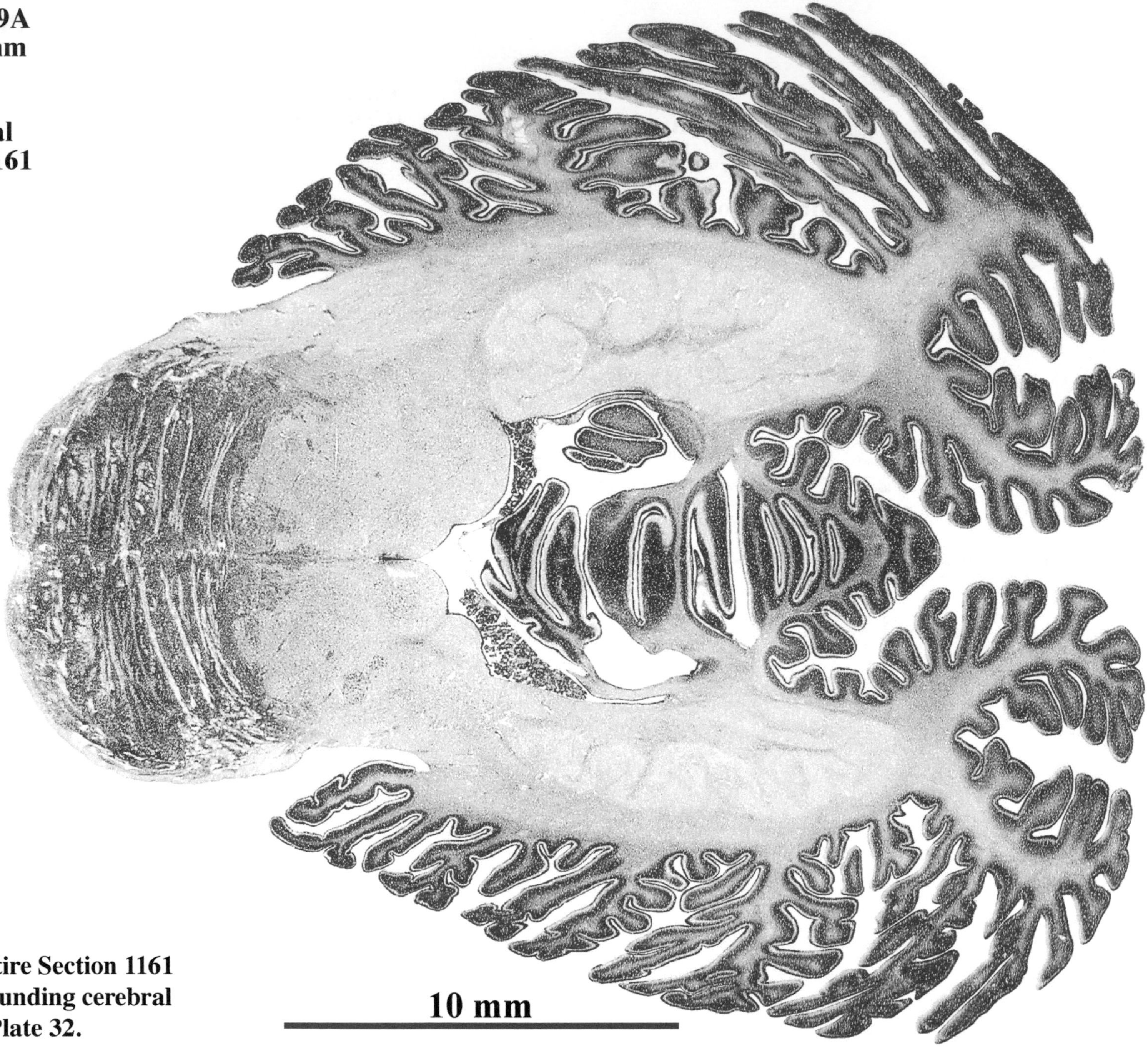

See the entire Section 1161 with surrounding cerebral cortex in Plate 32.

10 mm

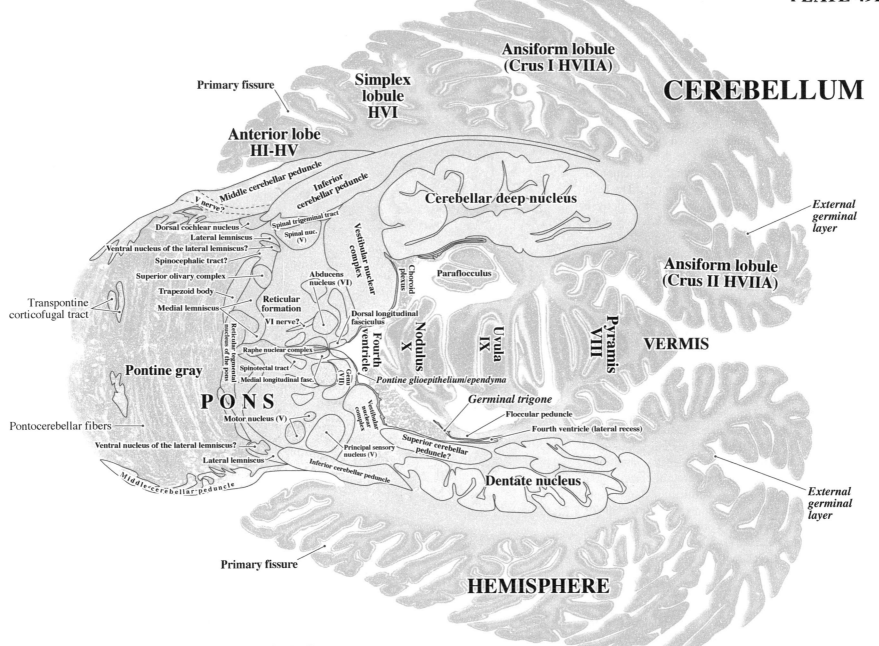

Ansiform lobule
(Crus I HVIIA)

CEREBELLUM

Primary fissure

Simplex
lobule
HVI

**Anterior lobe
HI-HV**

*External
germinal
layer*

Middle cerebellar peduncle

Inferior
cerebellar peduncle

Cerebellar deep nucleus

V nerve?

Dorsal cochlear nucleus

Spinal trigeminal tract

Lateral lemniscus

Spinal nuc.
(V)

Ventral nucleus of the lateral lemniscus?

Spinocephalic tract?

Vestibular nuclear
complex

**Ansiform lobule
(Crus II HVIIA)**

Superior olivary complex

Choroid
plexus

Paraflocculus

Abducens
nucleus (VI)

Trapezoid body

Transpontine
corticofugal tract

Medial lemniscus

**Reticular
formation**

Dorsal longitudinal
fasciculus

Pyramis
VIII

VI nerve?

Nodulus
X

Uvula
IX

VERMIS

Reticular tegmental
nucleus of the pons

Raphe nuclear complex

Fourth
ventricle

Pontine gray

Spinotectal tract

Genu
(VII)

Pontine glioepithelium/ependyma

P O N S

Medial longitudinal fasc.

Germinal trigone

Vestibular
nuclear
complex

Flocular peduncle

Motor nucleus (V)

Fourth ventricle (lateral recess)

Pontocerebellar fibers

Superior cerebellar
peduncle?

Ventral nucleus of the lateral lemniscus?

Principal sensory
nucleus (V)

Lateral lemniscus

Inferior cerebellar peduncle

Dentate nucleus

*External
germinal
layer*

Middle cerebellar peduncle

Primary fissure

HEMISPHERE

Germinal and transitional structures in *italics*

PART IV: Y217-65
CR 350 mm (GW 37)
Frontal

This specimen is case number W-217-65 (Perinatal RPSL) in the Yakovlev Collection. A male infant, weighing 3,057 grams, was born alive on August 10, 1965, and survived for 44 hours. Death occurred because a hyaline membrane obstructed the airway to the lungs. Upon autopsy, a subarachnoid hemmorhage in the left temporal/parietal area of the cerebral cortex was noted. The brain itself, cut in the frontal plane in 35-μm thick sections, is classified as a Normative Control in the Yakovlev Collection (Haleem, 1990). Although there is no photograph of this brain before it was embedded and cut, the photograph of the lateral view of another GW 37 brain that Larroche published in 1967 (**Figure 7**) is similar to the features of the brain in Y217-65. The approximate cutting plane of this brain is indicated in **Figure 8** (facing page) with lines superimposed on the GW 37 brain from the Larroche (1967) series. The dorsal part of each section is posterior to the ventral part. This brain is nearly even in mediolateral orientation. For example, the temporal poles on right vs. left sides of the brain appear at the same time in Section 611 (**Plate 54**). The sections chosen for illustration are more closely spaced to show small structures in the diencephalon, midbrain, pons, and medulla. Illustrated sections are spaced farther apart when they contain only large brain structures, such as the cerebral cortex, basal ganglia, and cerebellum. Low-magnification photographs of 21 Nissl-stained sections are shown in **Plates 50–70**. Different areas of the cerebral cortex are shown at very high magnification in **Plates 71–80**. The core of the brain and the cerebellum are shown at high magnification in **Plates 81-97**.

In the cortical regions of the telencephalon, remnants of the germinal matrices are present in all lobes of the cerebral cortex where the ***neuroepithelium/subventric-***

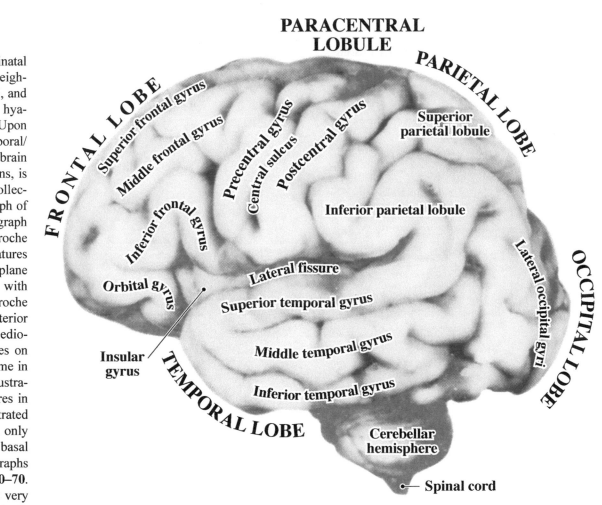

Figure 7. Lateral view of a GW 37 brain with major structures in the cerebral hemispheres labeled. (From the photographic series of J. C. Larroche (1967) Maturation morphologique du système nerveux central: ses rapports avec le développement pondéral du foetus et son age gestationnel. In: *Regional Development of the Brain in Early Life*, A. Minkowski (ed.), London: Blackwell, page 248.)

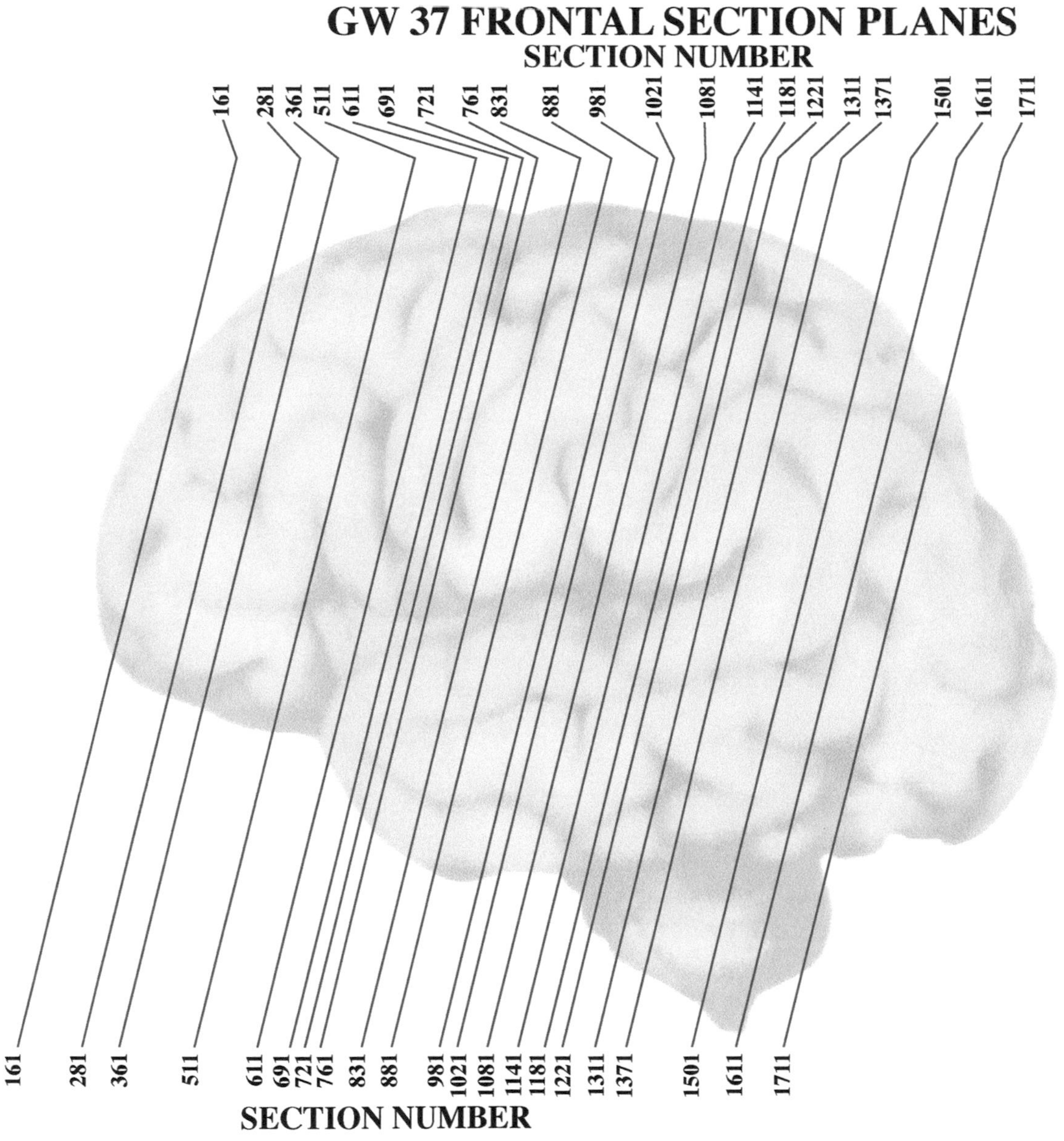

GW 37 FRONTAL SECTION PLANES
SECTION NUMBER

161 281 361 511 611 691 721 761 831 881 981 1021 1081 1141 1181 1221 1311 1371 1501 1611 1711

161 281 361 511 611 691 721 761 831 881 981 1021 1081 1141 1181 1221 1311 1371 1501 1611 1711

SECTION NUMBER

Figure 8. Lateral view of the same GW37 brain shown in **Figure 1** with the approximate locations and cutting angle of the sections of Y217-65. (From the photographic series of J. C. Larroche (1967) Maturation morphologique du système nerveux central: ses rapports avec le développement pondéral du foetus et son age gestationnel. In: *Regional Development of the Brain in Early Life*, A. Minkowski (ed.), London: Blackwell, page 248.)

ular zone is still generating neocortical interneurons. Remnants of migrating and sojourning neurons and/or glia are visible in all lobes of the cerebral cortex in the dwindling *stratified transitional fields*. Many neurons, glia, and their mitotic precursor cells are still migrating in the *rostral migratory stream* from a presumed source area in the germinal matrix at the junction between the cerebral cortex, striatum, and nucleus accumbens. Within the lateral parts of the cerebral cortex, streams of neurons and glia are in the *lateral migratory stream* that percolates through the claustrum, endopiriform nucleus, external capsule, and uncinate fasciculus. The cells appear to be heading toward the insular cortex, primary olfactory cortex, temporal cortex, and basolateral parts of the amygdaloid complex. There is active neurogenesis in the *subgranular zone* in the hilus of the dentate gyrus that is generating granule cells.

In the basal ganglia, there is a prominent *neuroepithelium/subventricular zone* overlying the striatum and nucleus accumbens where neurons are being generated. The striatal portion can be subdivided into anterolateral, anteromedial, and posterior parts. Other structures in the telencephalon, such as the septum, fornix, and Ammon's horn part of the hippocampus, have only a thin, darkly staining layer at the ventricle, and these are presumed to be generating glia, cells of the choroid plexus, and the ependymal lining of the ventricle

Most of the structures in the diencephalon appear to be settled and are maturing, and the third ventricle is lined by a thin *glioepithelium/ependyma*. In the midbrain and anterior pons, there is a slightly thicker and more convoluted *glioepithelium/ependyma* lining the posterior cerebral aqueduct and anterior fourth ventricle. The posterior pons and entire medulla have a thin *glioepithelium/ependyma* lining the rest of the fourth ventricle. The *external germinal layer* is prominent over the entire surface of the cerebellar cortex and is producing basket, stellate, and granule cells. The *germinal trigone* is visible at the base of the nodulus and along the floccular peduncle; choroid plexus cells and glia may be originating here.

PLATE 50A
CR 350 mm
GW 37
Y217-65
Coronal
Section 161

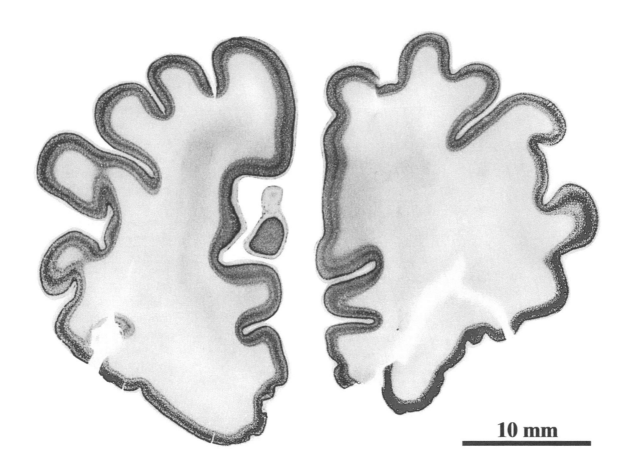

10 mm

Remnants of the germinal matrix
migratory streams, and transitional fields

1 *Subpial granular layer (cortical)*

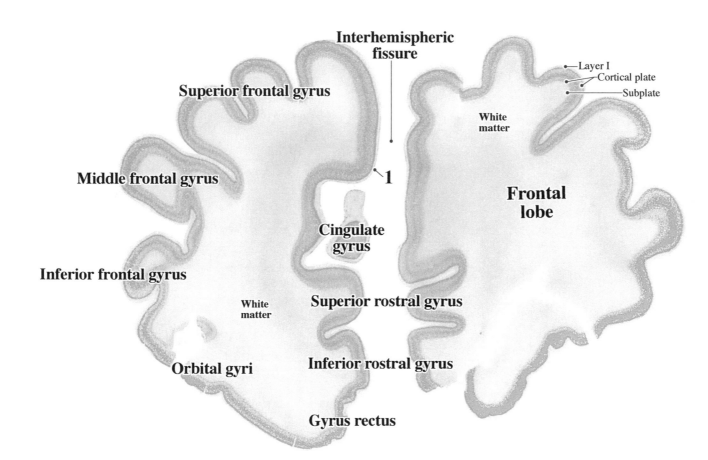

Interhemispheric fissure

Layer I
Cortical plate
Subplate

Superior frontal gyrus

White matter

Middle frontal gyrus

1

Frontal lobe

Cingulate gyrus

Inferior frontal gyrus

White matter

Superior rostral gyrus

Orbital gyri

Inferior rostral gyrus

Gyrus rectus

PLATE 51A
CR 350 mm
GW 37
Y217-65
Coronal
Section 281

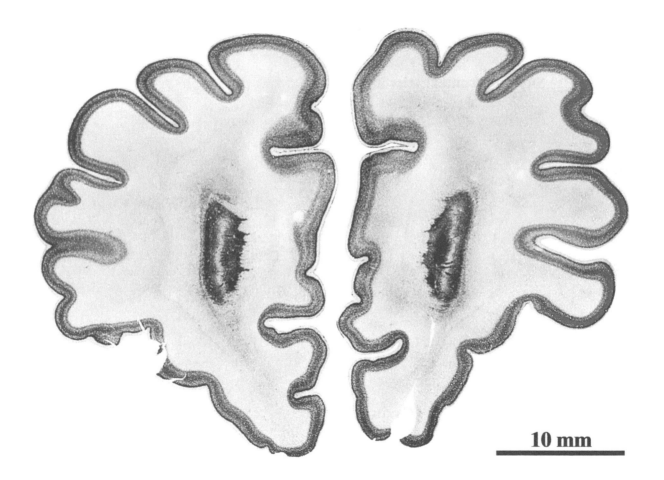

Remnants of the germinal matrix
migratory streams, and transitional fields

1 *Frontal neuroepithelium and subventricular zone*

2 *Frontal stratified transitional field*

3 *Subpial granular layer (cortical)*

10 mm

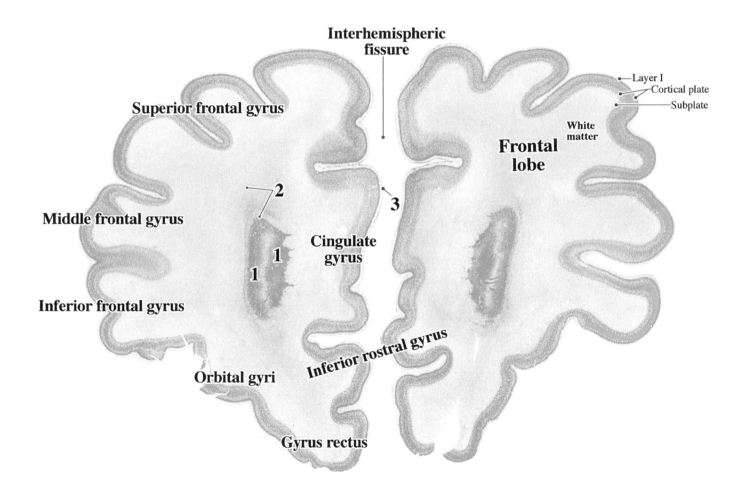

Interhemispheric
fissure

Superior frontal gyrus

Frontal
lobe

Layer I

Cortical plate

Subplate

White
matter

2

3

Middle frontal gyrus

1

Cingulate
gyrus

1

1

Inferior frontal gyrus

Inferior rostral gyrus

Orbital gyri

Gyrus rectus

PLATE 52A
CR 350 mm
GW 37
Y217-65
Coronal
Section 361

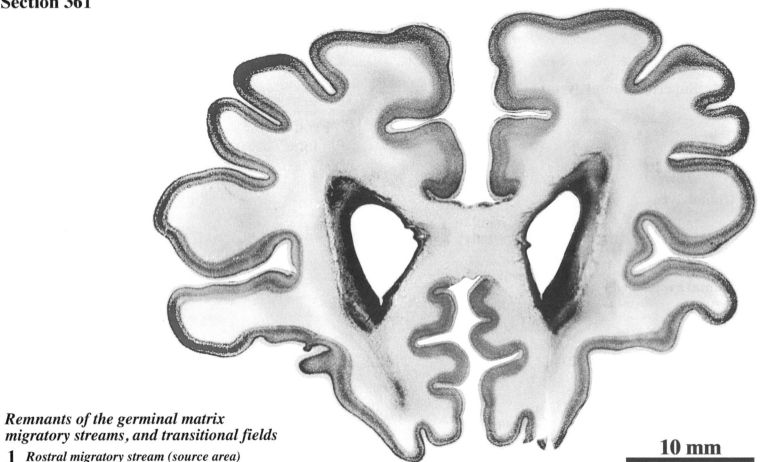

Remnants of the germinal matrix
migratory streams, and transitional fields

1 *Rostral migratory stream (source area)*

2 *Frontal NEP and SVZ*

3 *Frontal stratified transitional field*

4 *Callosal GEP*

5 *Anterolateral striatal NEP and SVZ*

6 *Subpial granular layer (cortical)*

10 mm

GEP - Glioepithelium
G/EP - Glioepithelium/ependyma
NEP - Neuroepithelium
STF - Stratified transitional field
SVZ - Subventricular zone

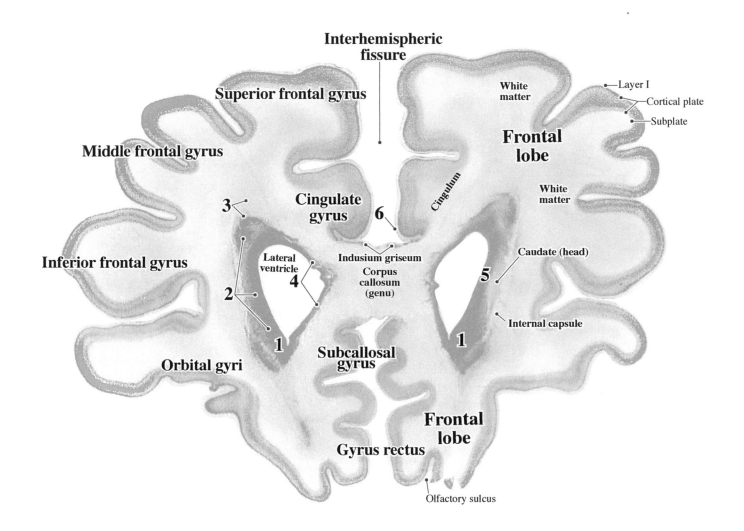

Interhemispheric fissure

Superior frontal gyrus

Middle frontal gyrus

Cingulate gyrus

Inferior frontal gyrus

Orbital gyri

Subcallosal gyrus

Gyrus rectus

White matter

Frontal lobe

Layer I

Cortical plate

Subplate

Cingulum

White matter

Caudate (head)

Internal capsule

Indusium griseum

Corpus callosum (genu)

Lateral ventricle

Frontal lobe

Olfactory sulcus

3

6

2

4

1

5

1

PLATE 53A
CR 350 mm
GW 37
Y217-65
Coronal
Section 511

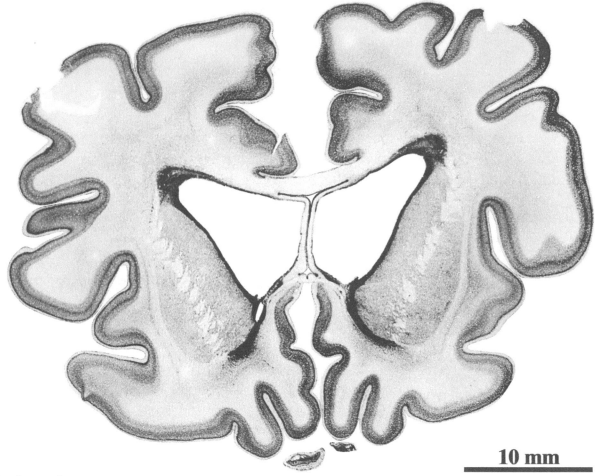

10 mm

Remnants of the germinal matrix
migratory streams, and transitional fields

1 *Rostral migratory stream*

2 *Frontal NEP and SVZ*

3 *Frontal STF*

4 *Frontal NEP and SVZ (intermingled with the source of the rostral migratory stream)*

5 *Callosal GEP*

6 *Callosal sling*

7 *Fornical GEP*

8 *Anterolateral striatal NEP and SVZ*

9 *Accumbent NEP and SVZ (intermingled with the source of the rostral migratory stream)*

10 *Subpial granular layer (cortical)*

GEP - Glioepithelium
NEP - Neuroepithelium
STF - Stratified transitional field
SVZ - Subventricular zone

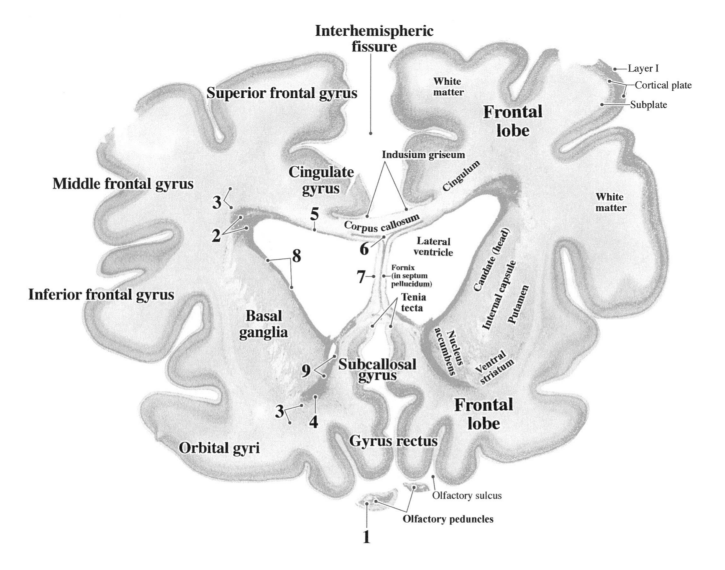

Interhemispheric fissure

Superior frontal gyrus

White matter

Frontal lobe

Layer I

Cortical plate

Subplate

Indusium griseum

Cingulate gyrus

Cingulum

Middle frontal gyrus

3

2

5

Corpus callosum

6

White matter

8

Lateral ventricle

Caudate (head)

Internal capsule

Putamen

Inferior frontal gyrus

7

Fornix (in septum pellucidum)

Tenia tecta

Basal ganglia

Nucleus accumbens

Ventral striatum

9

Subcallosal gyrus

Frontal lobe

3

4

Orbital gyri

Gyrus rectus

Olfactory sulcus

Olfactory peduncles

1

PLATE 54A
CR 350 mm
GW 37
Y217-65
Coronal
Section 611

See detail of the brain core
in Plates 81A and B.

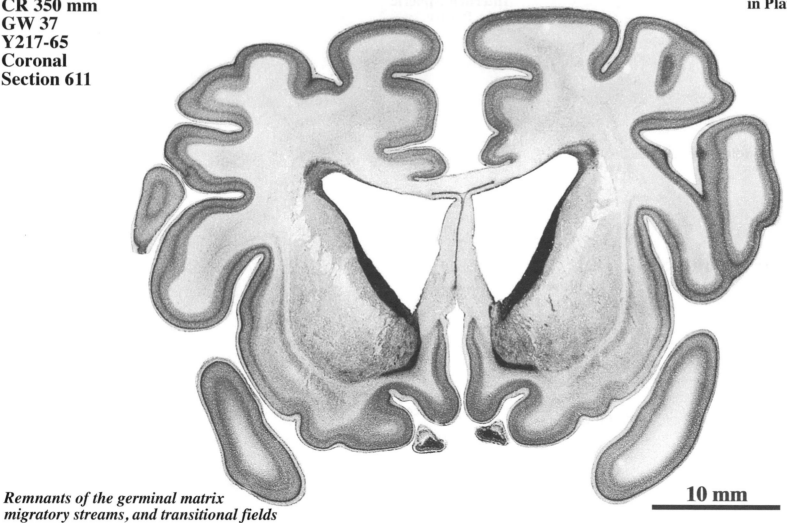

10 mm

Remnants of the germinal matrix
migratory streams, and transitional fields

1 *Rostral migratory stream*

2 *Frontal NEP and SVZ*

3 *Frontal STF*

4 *Callosal GEP*

5 *Callosal sling*

6 *Fornical GEP*

7 *Anterolateral striatal NEP and SVZ*

8 *Anteromedial striatal NEP and SVZ*

9 *Accumbent NEP and SVZ (intermingled with*
the source of the rostral migratory stream)

10 *Septal G/EP*

11 *Subpial granular layer (cortical)*

GEP - Glioepithelium
G/EP - Glioepithelium/ependyma
NEP - Neuroepithelium
STF - Stratified transitional field
SVZ - Subventricular zone

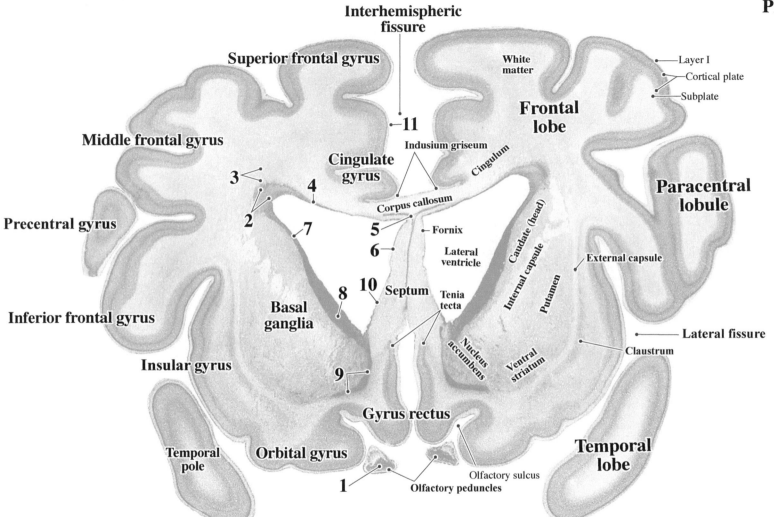

Interhemispheric fissure

Superior frontal gyrus

White matter

Layer I

Cortical plate

Subplate

Frontal lobe

Middle frontal gyrus

Indusium griseum

Cingulum

Cingulate gyrus

Paracentral lobule

3

4

Corpus callosum

Precentral gyrus

2

7

5

Fornix

6

Lateral ventricle

Caudate (head)

Internal capsule

External capsule

Inferior frontal gyrus

8

10

Septum

Tenia tecta

Putamen

Basal ganglia

Lateral fissure

Insular gyrus

Nucleus accumbens

Ventral striatum

Claustrum

9

Gyrus rectus

Temporal lobe

Temporal pole

Orbital gyrus

Olfactory sulcus

1

Olfactory peduncles

PLATE 55A
CR 350 mm
GW 37
Y217-65
Coronal
Section 691

See detail of the brain core
in Plates 82A and B.

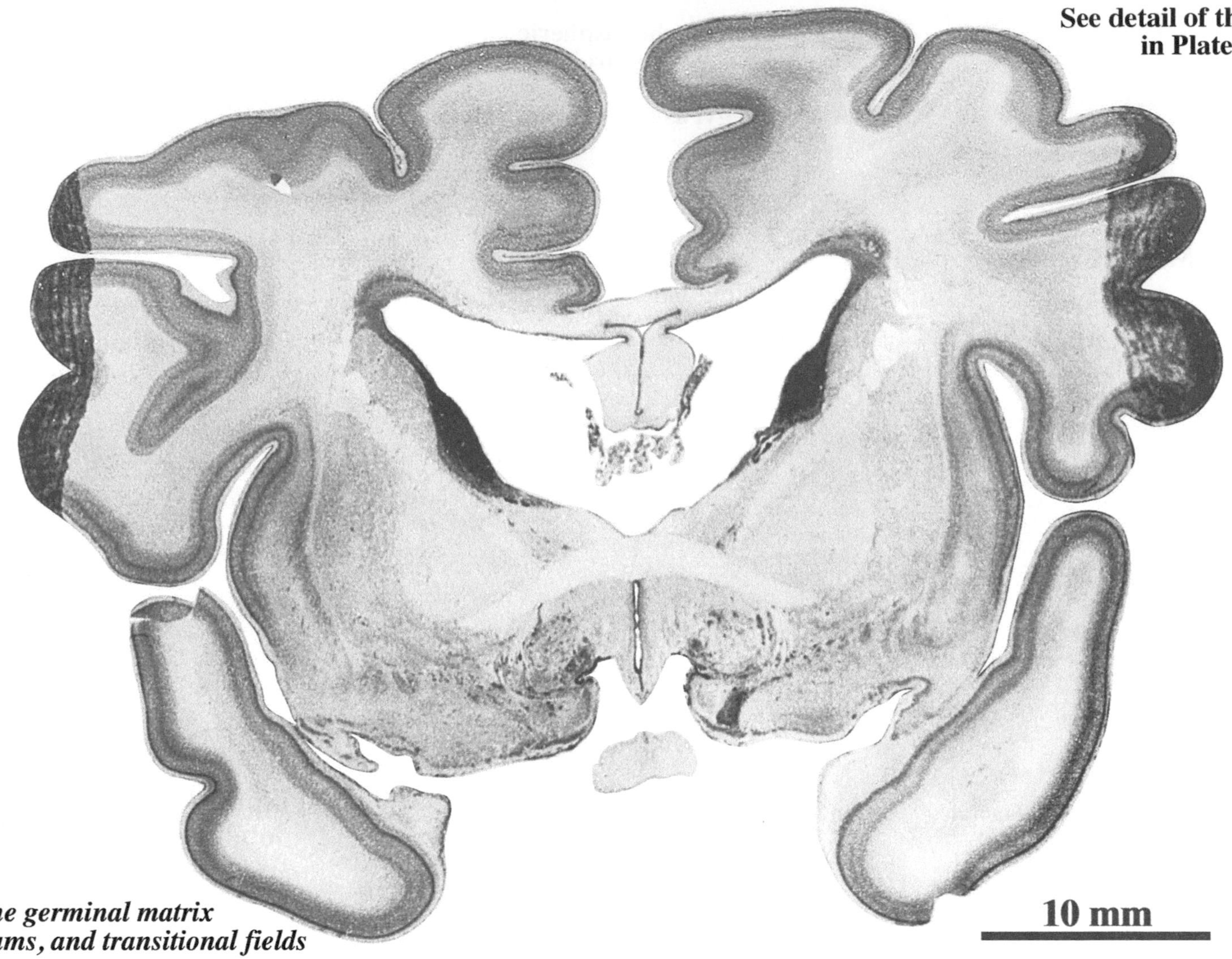

10 mm

Remnants of the germinal matrix
migratory streams, and transitional fields

1 *Rostral migratory stream*

2 *Frontal NEP and SVZ*

3 *Frontal STF*

4 *Callosal GEP*

5 *Callosal sling*

6 *Fornical GEP*

7 *Lateral migratory stream (cortical)*

8 *Anterolateral striatal NEP and SVZ*

9 *Anteromedial striatal NEP and SVZ*

10 *Strionuclear GEP*

11 *Diencephalic (preoptic) G/EP*

12 *Subpial granular layer (cortical)*

GEP - Glioepithelium
G/EP - Glioepithelium/ependyma
NEP - Neuroepithelium
STF - Stratified transitional field
SVZ - Subventricular zone

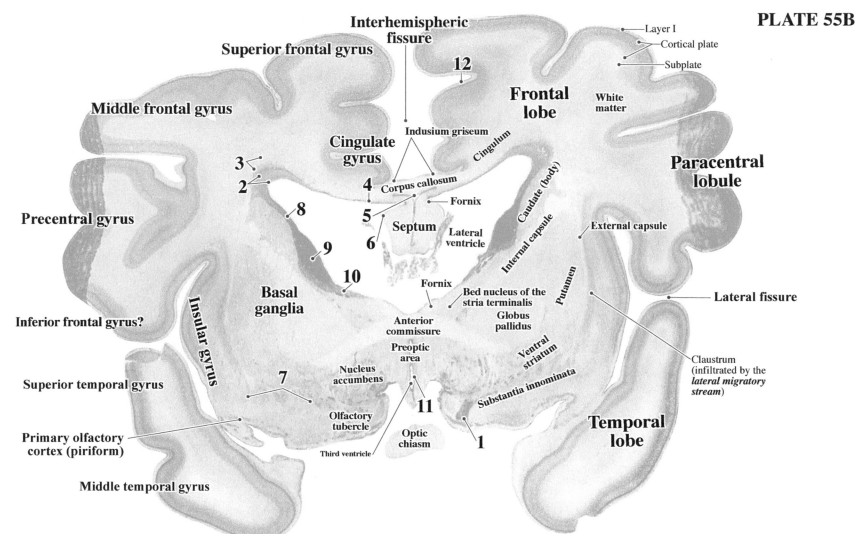

Interhemispheric fissure

Superior frontal gyrus

Middle frontal gyrus

Precentral gyrus

Inferior frontal gyrus?

Superior temporal gyrus

Primary olfactory cortex (piriform)

Middle temporal gyrus

Insular gyrus

Basal ganglia

Cingulate gyrus

Indusium griseum

Corpus callosum

Fornix

Septum

Lateral ventricle

Fornix

Anterior commissure

Preoptic area

Nucleus accumbens

Olfactory tubercle

Third ventricle

Optic chiasm

Bed nucleus of the stria terminalis

Globus pallidus

Ventral striatum

Substantia innominata

Frontal lobe

Layer I

Cortical plate

Subplate

White matter

Cingulum

Caudate (body)

Internal capsule

Putamen

External capsule

Lateral fissure

Claustrum (infiltrated by the *lateral migratory stream*)

Paracentral lobule

Temporal lobe

12

3

2

8

4

5

9

6

10

7

11

1

PLATE 56A
CR 350 mm
GW 37
Y217-65
Coronal
Section 721

See detail of the brain core
in Plates 83A and B.

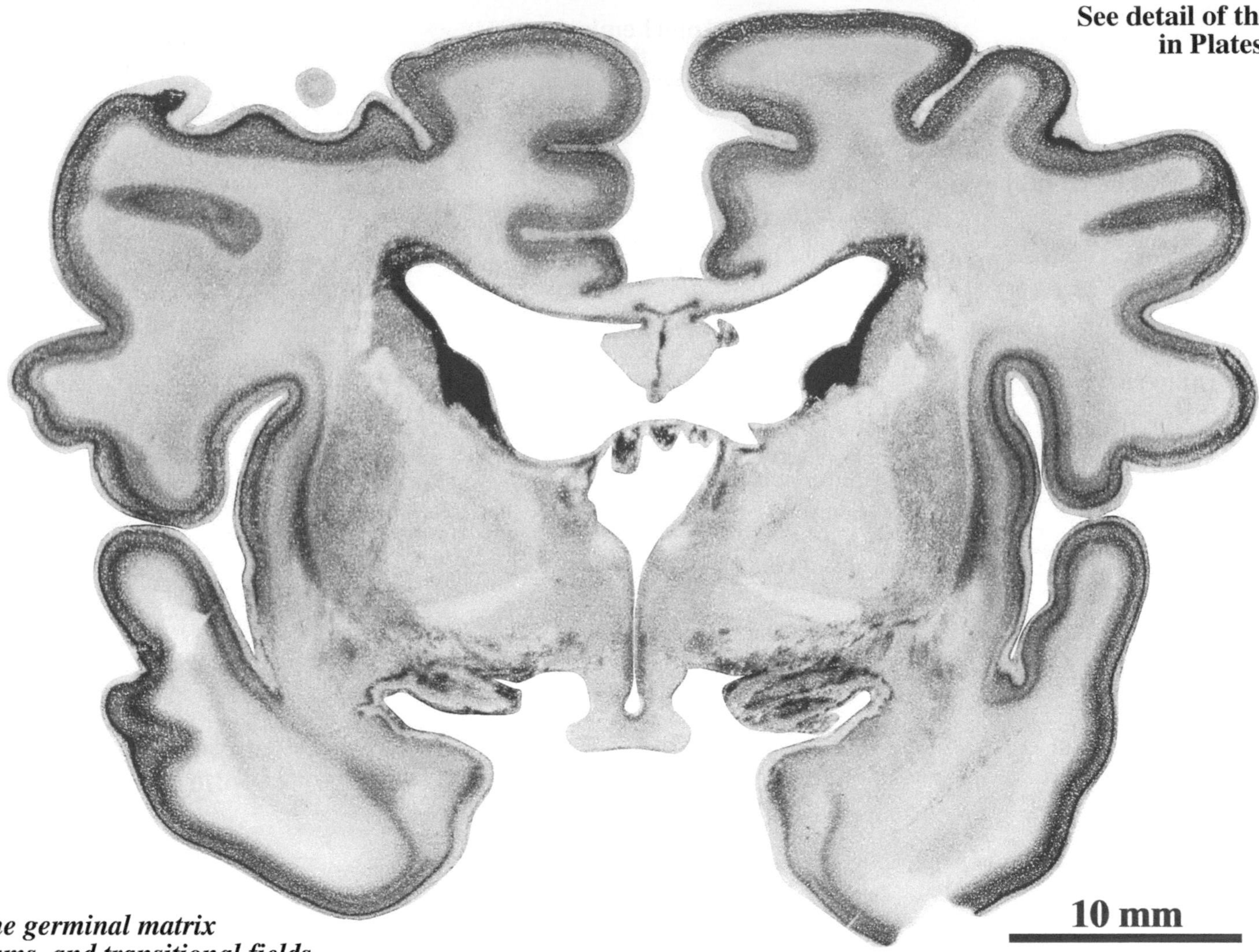

10 mm

Remnants of the germinal matrix
migratory streams, and transitional fields

1 *Frontal neuroepithelium and subventricular zone*

2 *Frontal stratified transitional field*

3 *Callosal glioepithelium*

4 *Callosal sling*

5 *Fornical glioepithelium*

6 *Lateral migratory stream (cortical)*

7 *Anterolateral striatal neuroepithelium and subventricular zone*

8 *Anteromedial striatal neuroepithelium and subventricular zone*

9 *Strionuclear glioepithelium*

10 *Diencephalic (epithalamic) glioepithelium/ependyma*

11 *Diencephalic (preoptic) glioepithelium/ependyma*

12 *Subpial granular layer (cortical)*

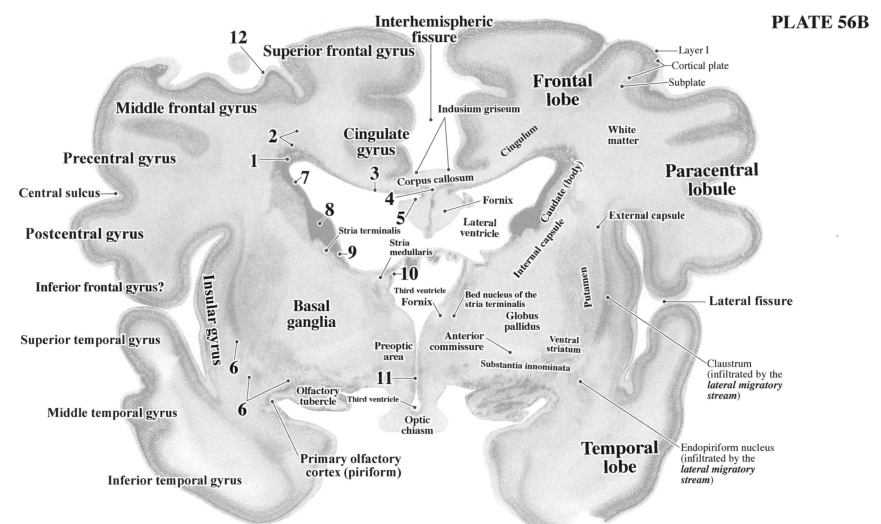

Interhemispheric fissure

Superior frontal gyrus

12

Middle frontal gyrus

Precentral gyrus

Central sulcus

Postcentral gyrus

Inferior frontal gyrus?

Superior temporal gyrus

Middle temporal gyrus

Inferior temporal gyrus

Insular gyrus

2

1

7

Cingulate gyrus

3

4

8

5

Stria terminalis

9

Stria medullaris

10

Third ventricle

Basal ganglia

6

6

Olfactory tubercle

Third ventricle

Preoptic area

11

Optic chiasm

Primary olfactory cortex (piriform)

Indusium griseum

Corpus callosum

Fornix

Lateral ventricle

Cingulum

Frontal lobe

Layer I

Cortical plate

Subplate

White matter

Caudate (body)

Paracentral lobule

External capsule

Internal capsule

Putamen

Bed nucleus of the stria terminalis

Fornix

Globus pallidus

Anterior commissure

Ventral striatum

Substantia innominata

Lateral fissure

Claustrum (infiltrated by the *lateral migratory stream*)

Endopiriform nucleus (infiltrated by the *lateral migratory stream*)

Temporal lobe

**PLATE 57A
CR 350 mm
GW 37
Y217-65
Coronal
Section 761**

✳ **See this area of cortex in Plates 71 and 72**

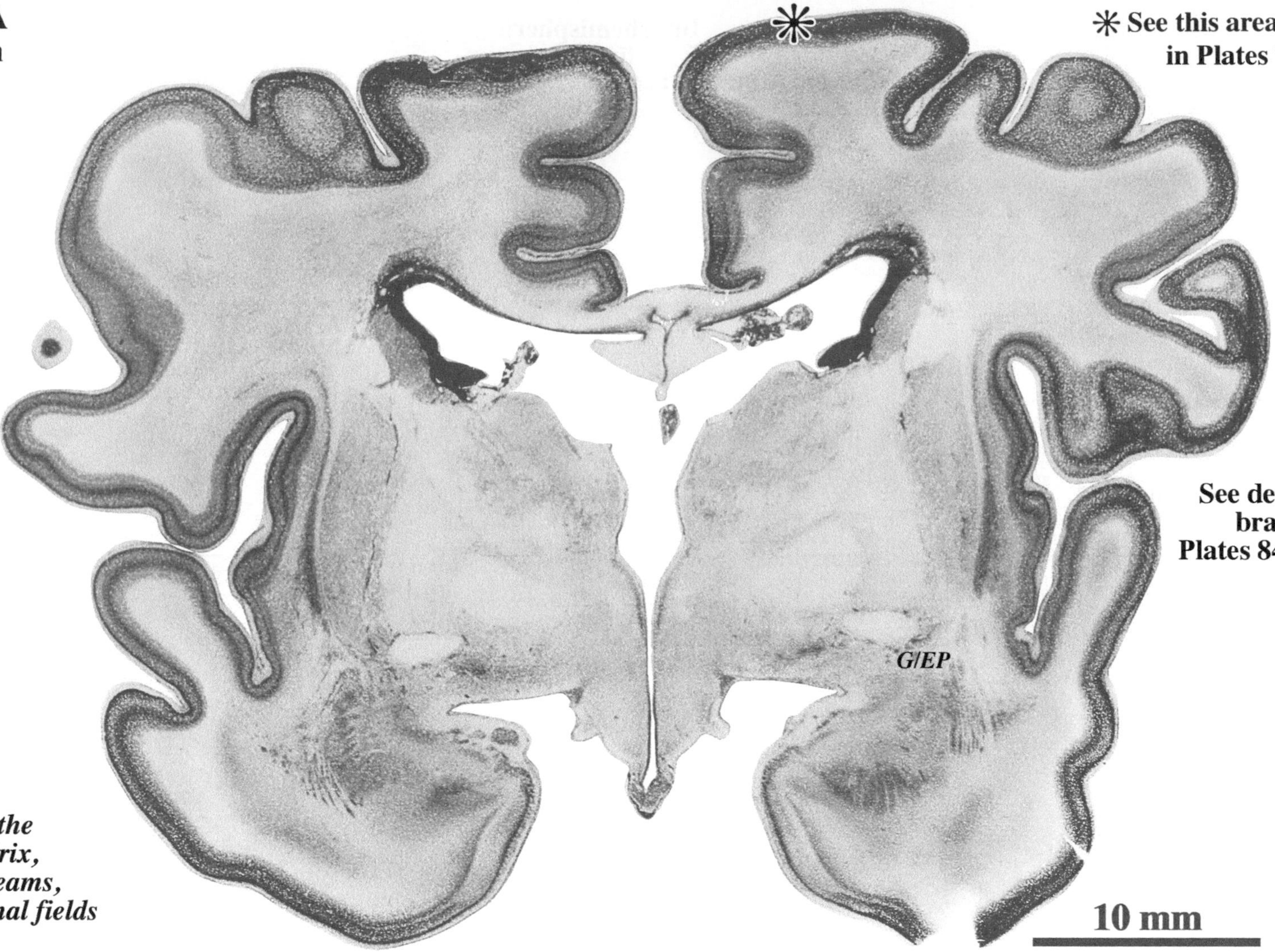

See detail of the brain core in Plates 84A and B.

G/EP

10 mm

Remnants of the germinal matrix, migratory streams, and transitional fields

1 *Frontal/paracentral NEP and SVZ*

2 *Frontal STF*

3 *Callosal GEP*

4 *Callosal sling*

5 *Fornical GEP*

6 *Lateral migratory stream (cortical)*

7 *Anterolateral striatal NEP and SVZ*

8 *Anteromedial striatal NEP and SVZ*

9 *Strionuclear GEP*

10 *Diencephalic (thalamic) G/EP*

11 *Diencephalic (hypothalamic) G/EP*

12 *Subpial granular layer (cortical)*

*GEP - Glioepithelium
NEP - Neuroepithelium
STF - Stratified transitional field
SVZ - Subventricular zone*

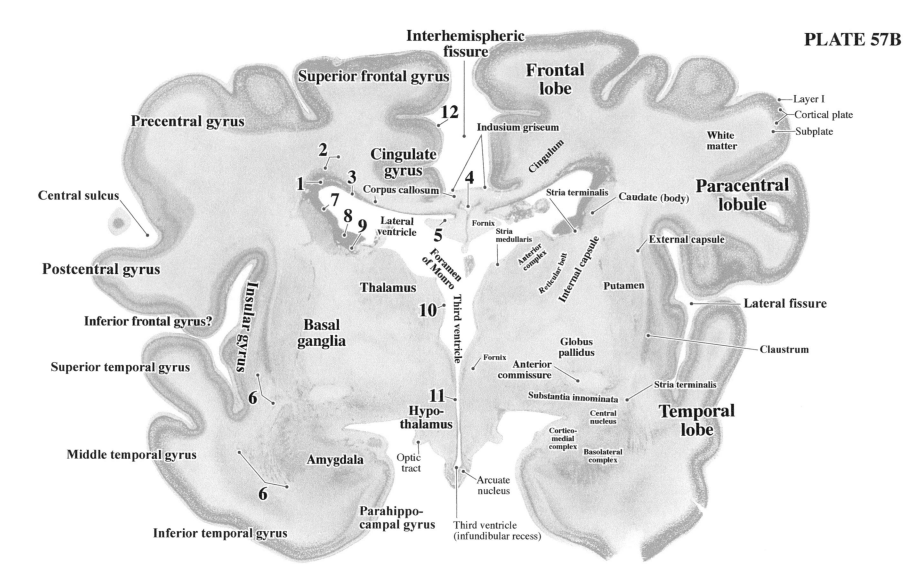

Interhemispheric fissure

Superior frontal gyrus

Frontal lobe

Precentral gyrus

12

Indusium griseum

Layer I
Cortical plate
Subplate

White matter

Cingulum

2

Cingulate gyrus

1

3

4

Corpus callosum

Stria terminalis

Caudate (body)

Paracentral lobule

Central sulcus

7

8 9

Lateral ventricle

5

Fornix

Stria medullaris

Anterior complex

External capsule

Reticular belt

Internal capsule

Postcentral gyrus

Insular gyrus

Thalamus

10

Foramen of Monro

Third ventricle

Putamen

Lateral fissure

Inferior frontal gyrus?

Basal ganglia

Claustrum

Superior temporal gyrus

6

Globus pallidus

Fornix

Anterior commissure

Stria terminalis

11

Substantia innominata

Temporal lobe

Middle temporal gyrus

6

Amygdala

Hypo-thalamus

Optic tract

Arcuate nucleus

Cortico-medial complex

Central nucleus

Basolateral complex

Inferior temporal gyrus

Parahippo-campal gyrus

Third ventricle (infundibular recess)

PLATE 58A
CR 350 mm
GW 37
Y217-65
Coronal
Section 831

See detail of the
brain core in
Plates 85A and B.

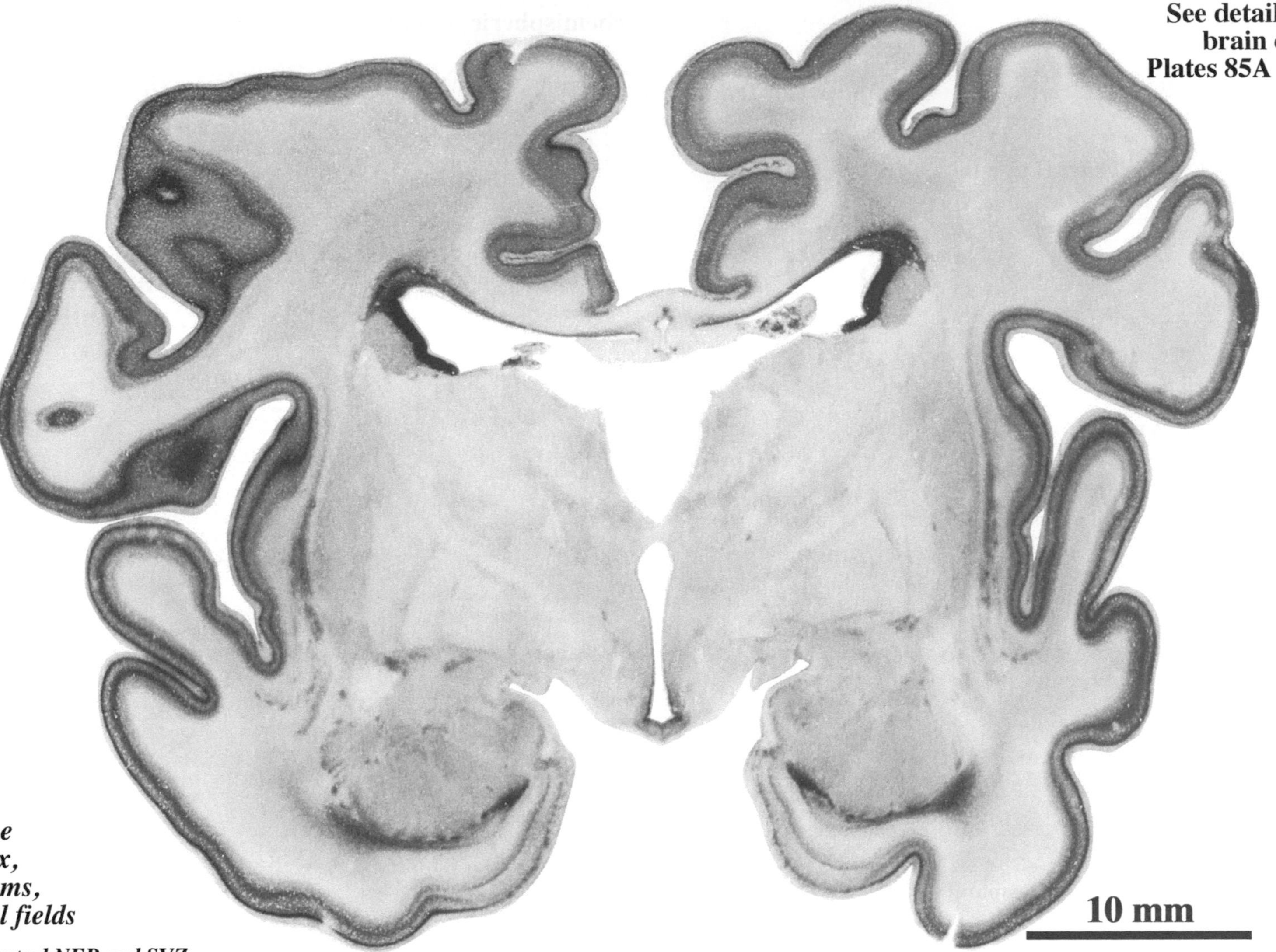

Remnants of the
germinal matrix,
migratory streams,
and transitional fields

1 *Frontal/paracentral NEP and SVZ*
2 *Frontal/paracentral STF*
3 *Callosal GEP*
4 *Callosal sling*
5 *Fornical GEP*
6 *Parahippocampal NEP, SVZ, and STF*
7 *Lateral migratory stream (cortical)*

8 *Amygdaloid G/EP*
9 *Posterior striatal NEP and SVZ*
10 *Strionuclear GEP*
11 *Diencephalic (thalamic) G/EP*
12 *Diencephalic (hypothalamic) G/EP*
13 *Subpial granular layer*

10 mm

GEP - Glioepithelium
G/EP - Glioepithelium/ependyma
NEP - Neuroepithelium
STF - Stratified transitional field
SVZ - Subventricular zone

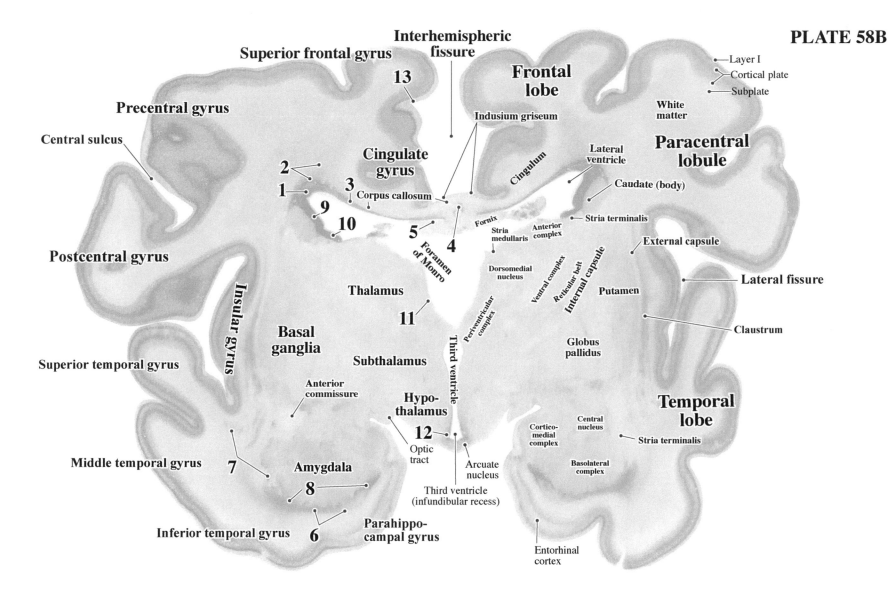

Interhemispheric fissure

Superior frontal gyrus

13

Frontal lobe

Layer I
Cortical plate
Subplate

Precentral gyrus

Indusium griseum

White matter

Central sulcus

Paracentral lobule

Cingulum

Lateral ventricle

Cingulate gyrus

Caudate (body)

2

1

3

Corpus callosum

9

10

Fornix

Stria terminalis

5

Stria medullaris

Anterior complex

External capsule

4

Postcentral gyrus

Foramen of Monro

Dorsomedial nucleus

Ventral complex

Reticular belt

Internal capsule

Lateral fissure

Thalamus

Putamen

Insular gyrus

Periventricular complex

11

Claustrum

Basal ganglia

Subthalamus

Globus pallidus

Superior temporal gyrus

Third ventricle

Temporal lobe

Anterior commissure

Hypo-thalamus

Central nucleus

Stria terminalis

12

Cortico-medial complex

Middle temporal gyrus

7

Amygdala

Optic tract

Basolateral complex

Arcuate nucleus

8

Inferior temporal gyrus

6

Parahippo-campal gyrus

Third ventricle (infundibular recess)

Entorhinal cortex

PLATE 59A
CR 350 mm
GW 37
Y217-65
Coronal
Section 881

See detail of the
brain core in
Plates 86A and B.

Remnants of the
germinal matrix,
migratory streams,
and transitional fields

1 *Frontal/paracentral NEP and SVZ*

2 *Frontal/paracentral STF*

3 *Callosal GEP*

4 *Callosal sling*

5 *Fornical GEP*

6 *Parahippocampal NEP, SVZ, and STF*

7 *Temporal NEP and SVZ*

8 *Temporal STF*

9 *Alvear GEP*

10 *Lateral migratory stream (cortical)*

11 *Amygdaloid G/EP*

12 *Posterior striatal NEP and SVZ*

13 *Strionuclear GEP*

14 *Diencephalic (thalamic) G/EP*

15 *Diencephalic (hypothalamic) G/EP*

16 *Subpial granular layer*

GEP - Glioepithelium
G/EP - Glioepithelium/ependyma
NEP - Neuroepithelium
STF - Stratified transitional field
SVZ - Subventricular zone

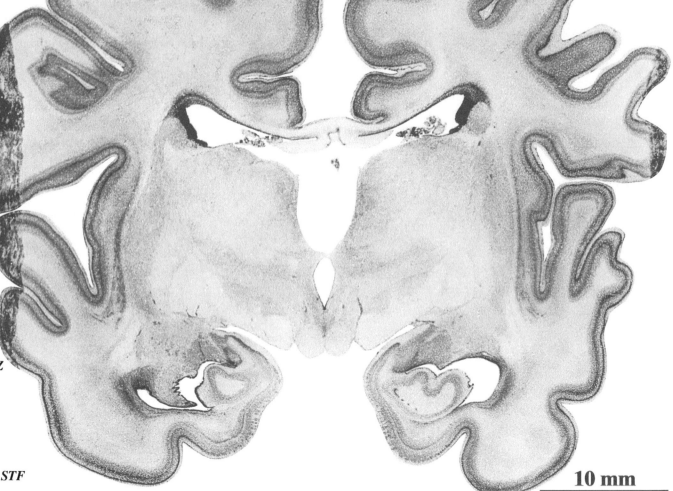

10 mm

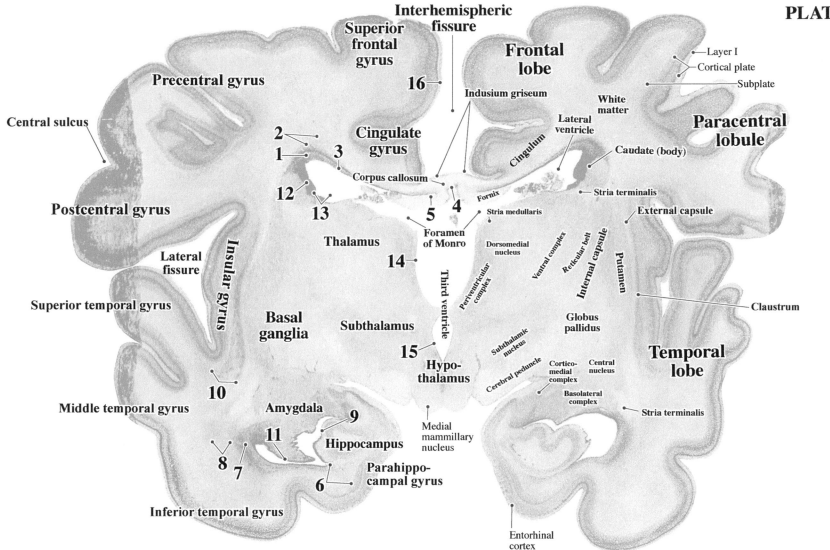

Interhemispheric
fissure

Superior
frontal
gyrus

Frontal
lobe

Precentral gyrus

16

Indusium griseum

Layer I

Cortical plate

Subplate

White
matter

Central sulcus

Cingulate
gyrus

Lateral
ventricle

Cingulum

Paracentral
lobule

2

1

3

Corpus callosum

Caudate (body)

Postcentral gyrus

12

13

Fornix

Stria terminalis

4

External capsule

5

Stria medullaris

Lateral
fissure

Insular gyrus

Thalamus

Foramen
of Monro

Dorsomedial
nucleus

Ventral complex

Reticular belt

Internal capsule

Putamen

Superior temporal gyrus

14

Third ventricle

Periventricular
complex

Claustrum

Basal
ganglia

Subthalamus

Globus
pallidus

Middle temporal gyrus

Subthalamic
nucleus

Temporal
lobe

10

15

Hypo-
thalamus

Cerebral peduncle

Cortico-
medial
complex

Central
nucleus

Amygdala

9

Basolateral
complex

Stria terminalis

11

Hippocampus

8   7

Medial
mammillary
nucleus

6

Parahippo-
campal gyrus

Inferior temporal gyrus

Entorhinal
cortex

PLATE 60A
CR 350 mm
GW 37
Y217-65
Coronal
Section 981

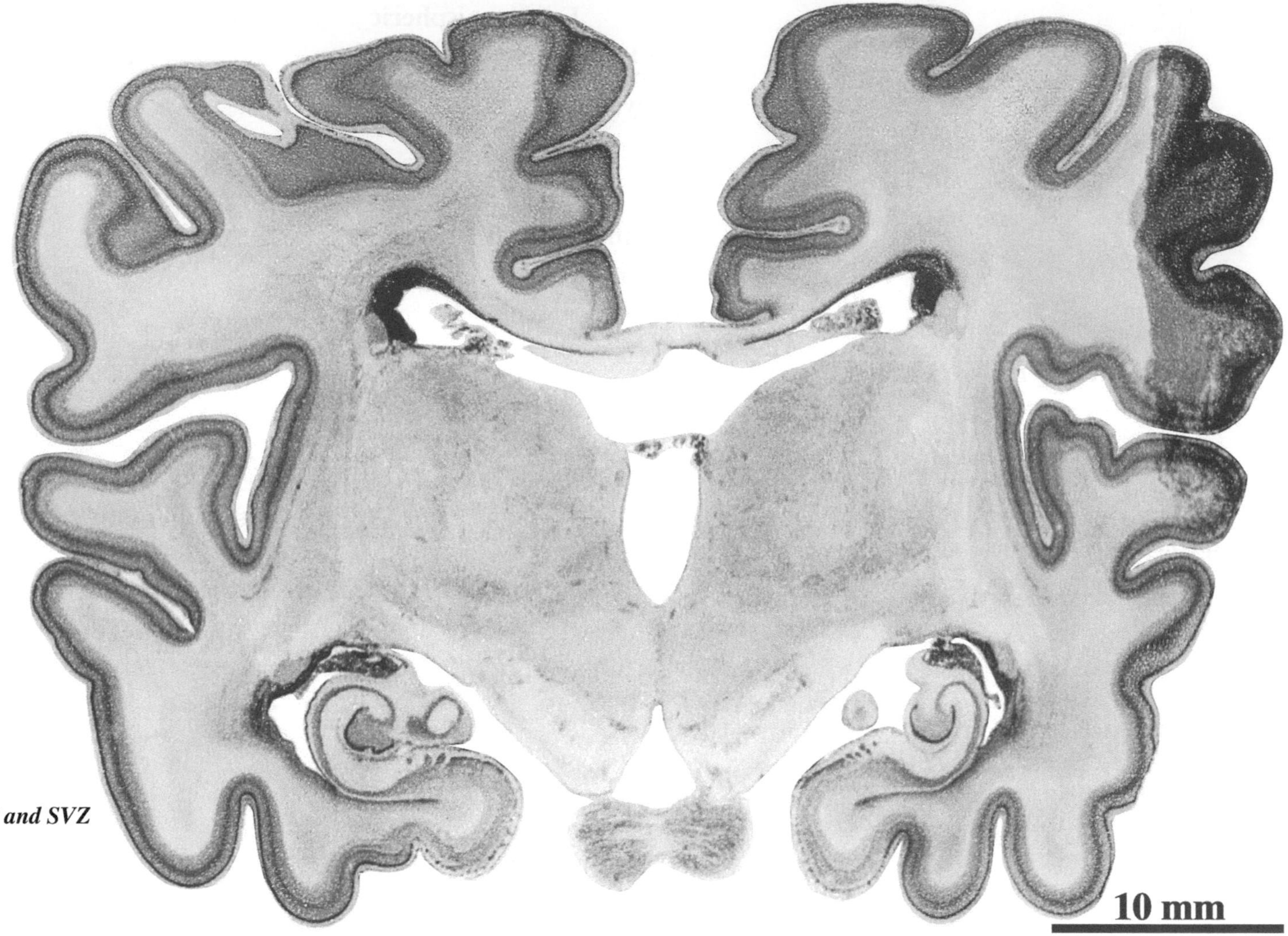

See detail of the
brain core in
Plates 87A and B.

Remnants of the
germinal matrix,
migratory streams,
and transitional fields

1 *Frontal/paracentral NEP and SVZ*

2 *Frontal/paracentral STF*

3 *Callosal GEP*

4 *Callosal sling*

5 *Fornical GEP*

6 *Parahippocampal NEP, SVZ, and STF*

7 *Temporal NEP and SVZ*

8 *Temporal STF*

9 *Alvear GEP*

10 *Subgranular zone (dentate)*

11 *Lateral migratory stream (cortical)*

12 *Posterior striatal NEP and SVZ*

13 *Strionuclear GEP*

14 *Diencephalic (thalamic) G/EP*

15 *Subpial granular layer (cortical)*

GEP - Glioepithelium
G/EP - Glioepithelium/ependyma
NEP - Neuroepithelium
STF - Stratified transitional field
SVZ - Subventricular zone

10 mm

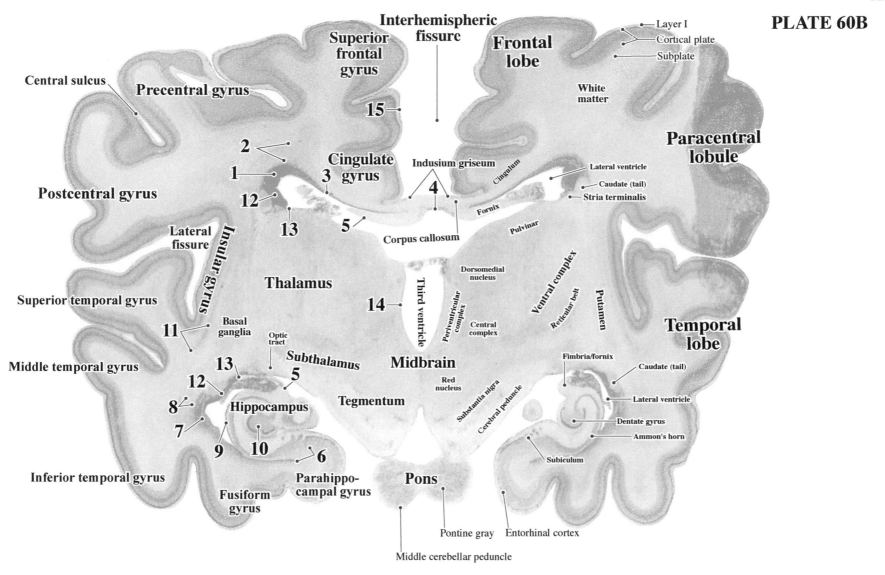

Interhemispheric fissure

Superior frontal gyrus

Frontal lobe

Layer I
Cortical plate
Subplate

Central sulcus

Precentral gyrus

15

White matter

Paracentral lobule

2

1

Cingulate gyrus

Indusium griseum

Cingulum

Lateral ventricle

12

3

4

Caudate (tail)

Stria terminalis

Postcentral gyrus

13

5

Fornix

Corpus callosum

Pulvinar

Lateral fissure

Insular gyrus

Dorsomedial nucleus

Ventral complex

Thalamus

Third ventricle

Periventricular complex

Reticular belt

Putamen

14

Superior temporal gyrus

11

Basal ganglia

Central complex

Temporal lobe

Optic tract

Middle temporal gyrus

13

Subthalamus

Midbrain

Fimbria/fornix

Caudate (tail)

12

Red nucleus

5

Substantia nigra

Lateral ventricle

8

Hippocampus

Tegmentum

Cerebral peduncle

Dentate gyrus

7

Ammon's horn

9

10

6

Subiculum

Inferior temporal gyrus

Parahippo-campal gyrus

Pons

Fusiform gyrus

Pontine gray

Entorhinal cortex

Middle cerebellar peduncle

PLATE 61A
CR 350 mm
GW 37
Y217-65
Coronal
Section 1021

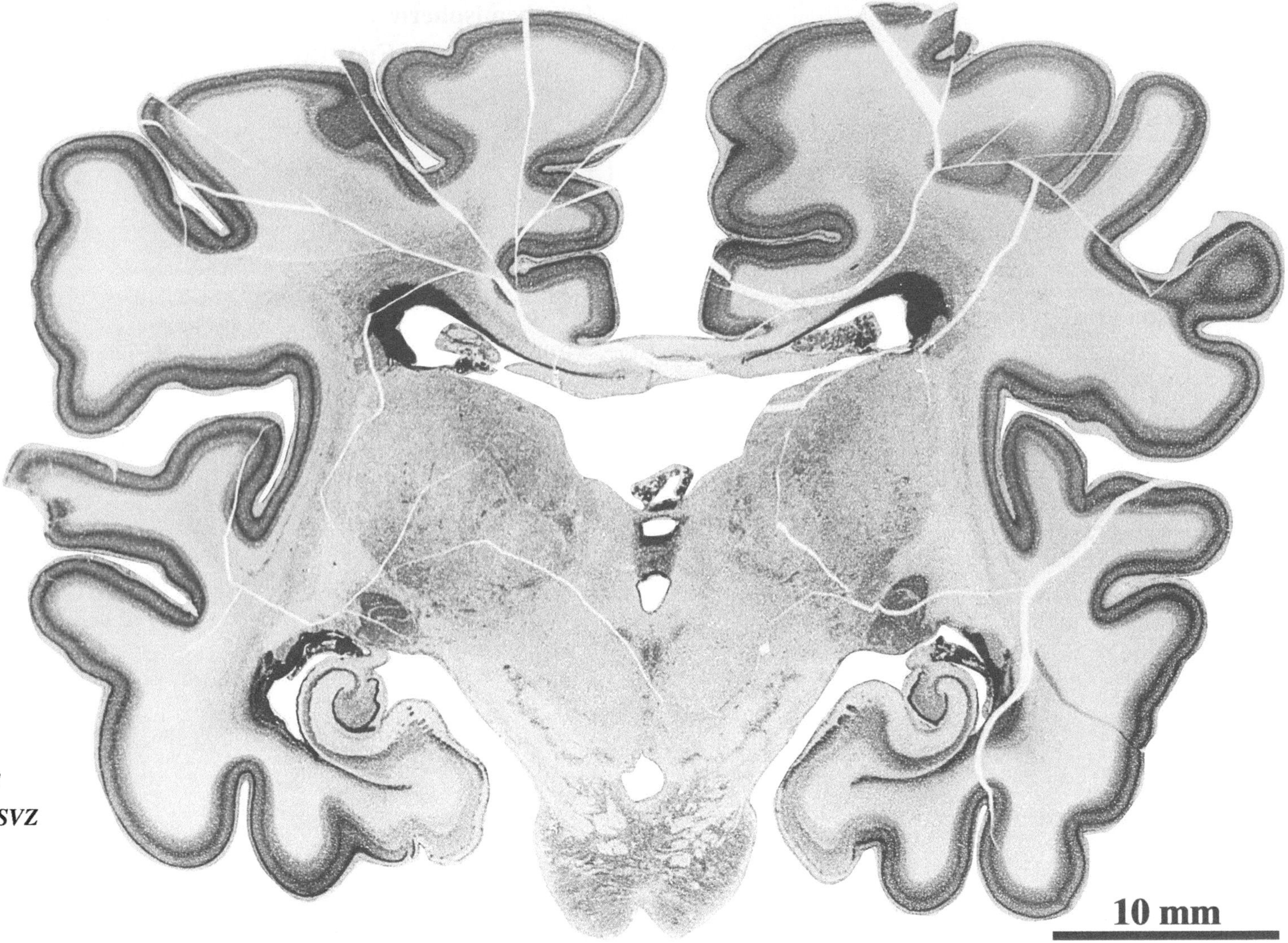

See detail of the brain core in Plates 88A and B.

Remnants of the germinal matrix, migratory streams, and transitional fields

1 *Paracentral NEP and SVZ*

2 *Paracentral STF*

3 *Callosal GEP*

4 *Callosal sling*

5 *Fornical GEP*

6 *Parahippocampal NEP, SVZ, and STF*

7 *Temporal NEP and SVZ*

8 *Temporal STF*

9 *Alvear GEP*

10 *Subgranular zone (dentate)*

11 *Lateral migratory stream (cortical)*

12 *Posterior striatal NEP and SVZ*

13 *Strionuclear GEP*

14 *Mesencephalic G/EP*

15 *Subpial granular layer (cortical)*

GEP - Glioepithelium
G/EP - Glioepithelium/ependyma
NEP - Neuroepithelium
STF - Stratified transitional field
SVZ - Subventricular zone

10 mm

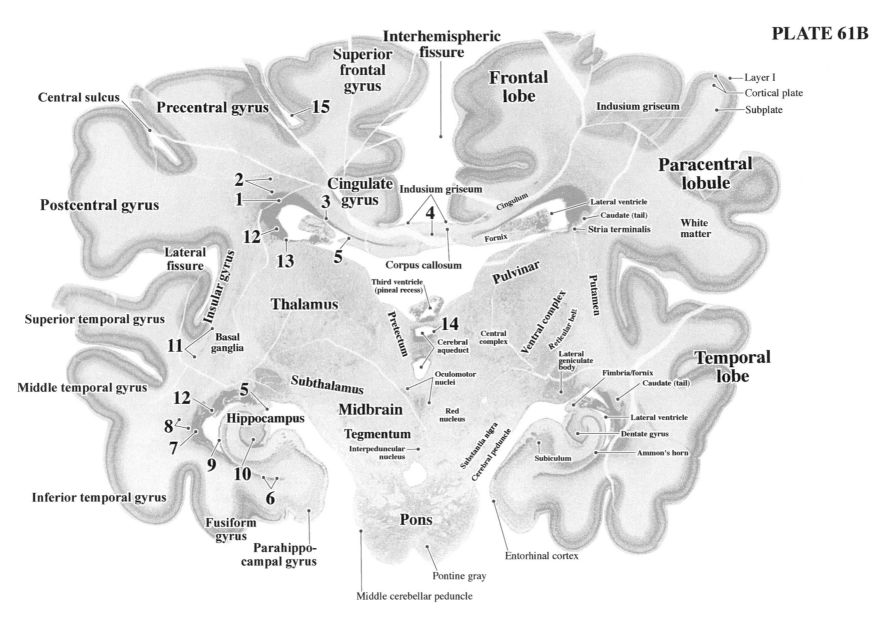

Interhemispheric fissure

Superior frontal gyrus

Frontal lobe

Layer I
Cortical plate
Subplate

Central sulcus

Precentral gyrus

15

Indusium griseum

Paracentral lobule

Postcentral gyrus

2
1

3

Cingulate gyrus

Indusium griseum

Cingulum

Lateral ventricle
Caudate (tail)
Stria terminalis

White matter

4

Fornix

12

13

5

Corpus callosum

Pulvinar

Putamen

Lateral fissure

Insular gyrus

Third ventricle (pineal recess)

Superior temporal gyrus

Basal ganglia

11

Thalamus

Pretectum

14

Cerebral aqueduct

Central complex

Ventral complex

Reticular belt

Lateral geniculate body

Temporal lobe

Middle temporal gyrus

Subthalamus

Oculomotor nuclei

Fimbria/fornix
Caudate (tail)

12

5

Midbrain

Red nucleus

Lateral ventricle

8

Hippocampus

Tegmentum

Dentate gyrus

7

Interpeduncular nucleus

Substantia nigra

Ammon's horn

9

10

6

Subiculum

Inferior temporal gyrus

Cerebral peduncle

Pons

Fusiform gyrus

Parahippo-campal gyrus

Entorhinal cortex

Pontine gray

Middle cerebellar peduncle

PLATE 62A
CR 350 mm
GW 37
Y217-65
Coronal
Section 1081

✳ See this area of cortex in
Plates 73 and 74.

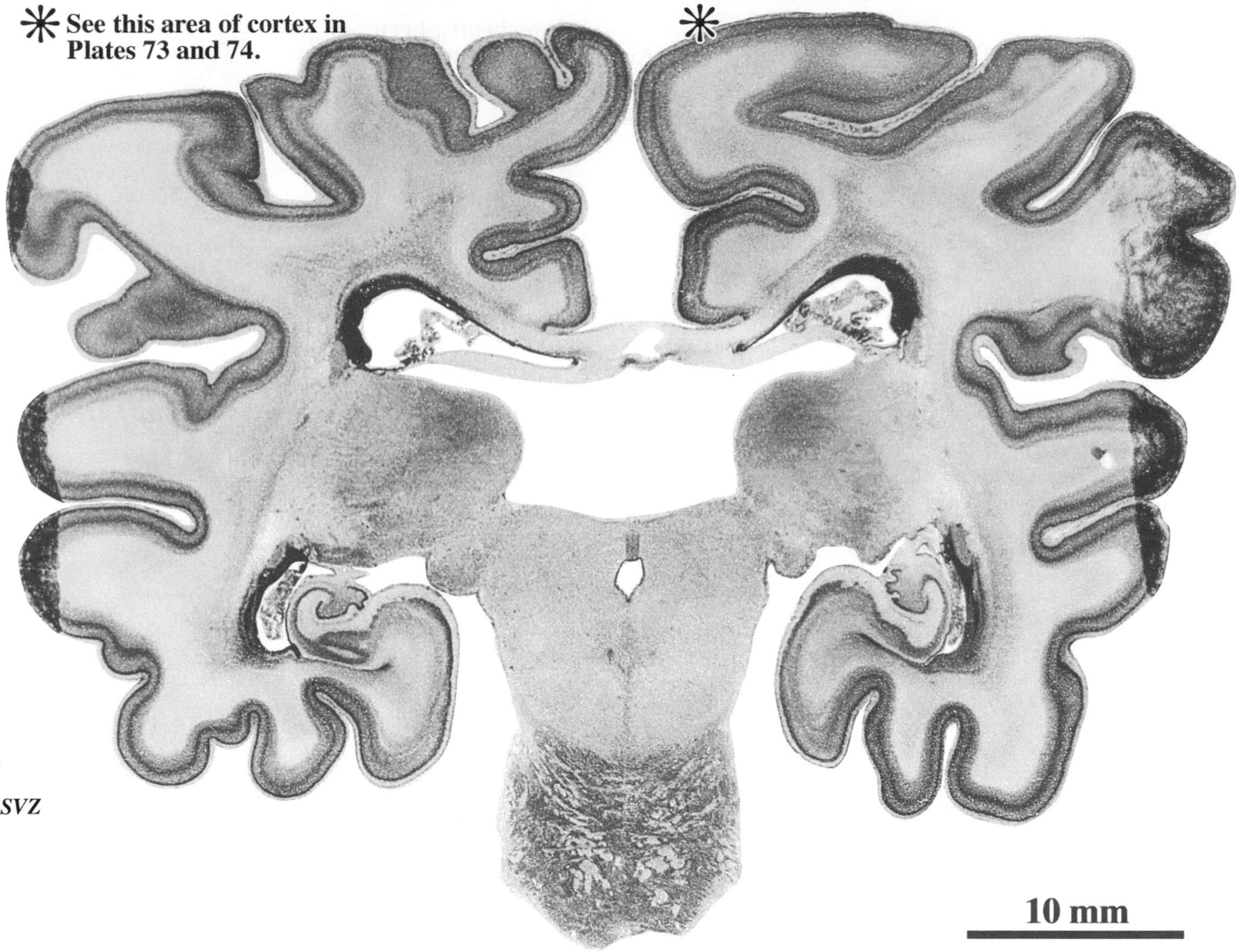

✳

See detail of the
brain core in
Plates 89A and B.

Remnants of the
germinal matrix,
migratory streams,
and transitional fields

1 *Paracentral NEP and SVZ*

2 *Paracentral STF*

3 *Callosal GEP*

4 *Callosal sling*

5 *Fornical GEP*

6 *Parahippocampal NEP, SVZ, and STF*

7 *Temporal NEP and SVZ*

8 *Temporal STF*

9 *Alvear GEP*

10 *Subgranular zone (dentate)*

11 *Posterior striatal NEP and SVZ*

12 *Strionuclear GEP*

13 *Mesencephalic G/EP*

14 *Subpial granular layer (cortical)*

10 mm

GEP - Glioepithelium
G/EP - Glioepithelium/ependyma
NEP - Neuroepithelium
STF - Stratified transitional field
SVZ - Subventricular zone

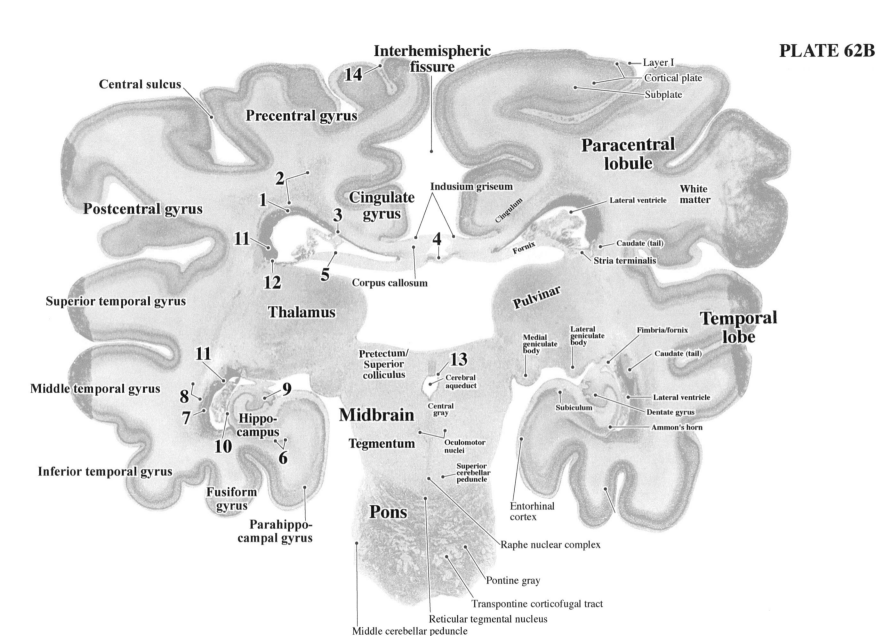

Interhemispheric
fissure

Central sulcus

Precentral gyrus

14

Layer I
Cortical plate
Subplate

**Paracentral
lobule**

Postcentral gyrus

2

1

Cingulate
gyrus

3

4

Indusium griseum

Cingulum

White
matter

Lateral ventricle

11

Fornix

Caudate (tail)

Stria terminalis

12

5

Corpus callosum

Thalamus

Pulvinar

**Temporal
lobe**

Superior temporal gyrus

11

Pretectum/
Superior
colliculus

13

Medial
geniculate
body

Lateral
geniculate
body

Fimbria/fornix

Caudate (tail)

Cerebral
aqueduct

9

8

Central
gray

Subiculum

Lateral ventricle

7

Dentate gyrus

Middle temporal gyrus

Hippo-
campus

6

Midbrain

Ammon's horn

10

Tegmentum

Oculomotor
nuclei

Inferior temporal gyrus

Superior
cerebellar
peduncle

Entorhinal
cortex

Fusiform
gyrus

Pons

Parahippo-
campal gyrus

Raphe nuclear complex

Pontine gray

Transpontine corticofugal tract

Reticular tegmental nucleus

Middle cerebellar peduncle

PLATE 63A
CR 350 mm
GW 37
Y217-65
Coronal
Section 1141

See detail of the brain core in Plates 90A and B.

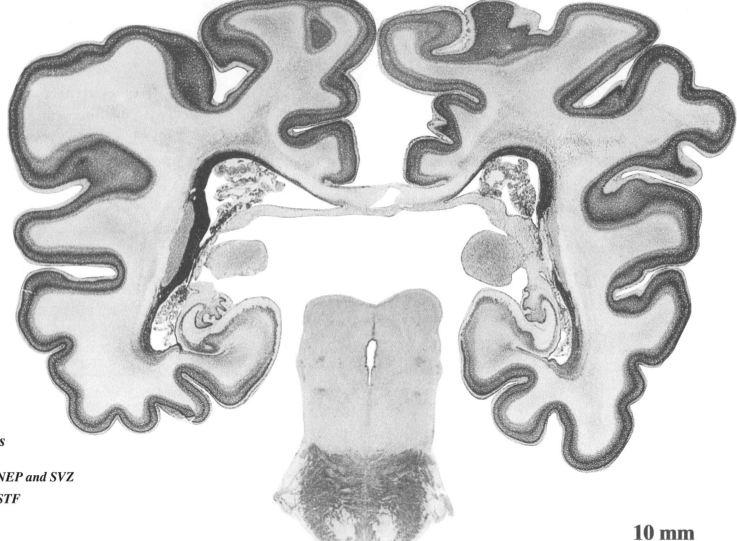

Remnants of the germinal matrix, migratory streams, and transitional fields

1 *Paracentral/parietal NEP and SVZ*

2 *Paracentral/parietal STF*

3 *Callosal GEP*

4 *Fornical GEP*

5 *Parahippocampal NEP, SVZ, and STF*

6 *Temporal NEP and SVZ*

7 *Temporal STF*

8 *Alvear GEP*

9 *Subgranular zone (dentate)*

10 *Posterior striatal NEP and SVZ*

11 *Strionuclear GEP*

12 *Mesencephalic G/EP*

13 *Subpial granular layer (cortical)*

GEP - Glioepithelium
G/EP - Glioepithelium/ependyma
NEP - Neuroepithelium
STF - Stratified transitional field
SVZ - Subventricular zone

10 mm

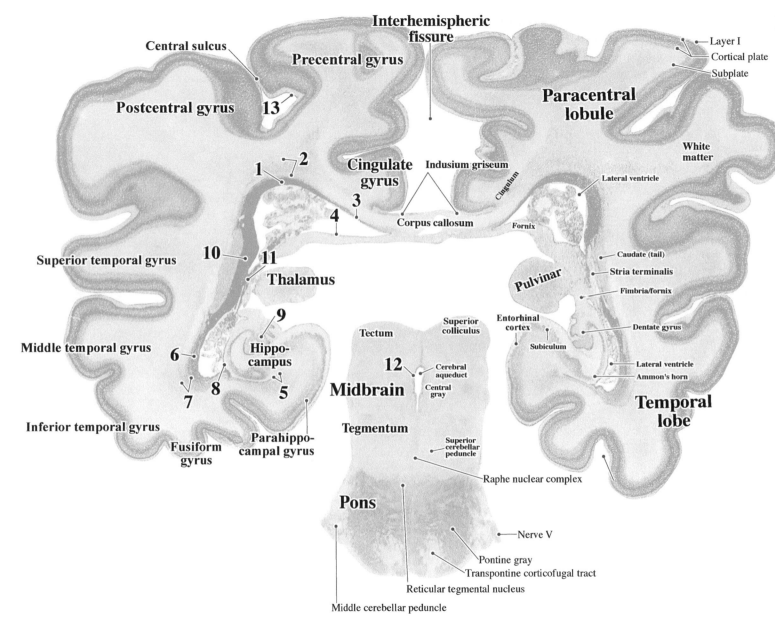

Interhemispheric fissure

Central sulcus

Precentral gyrus

Layer I
Cortical plate
Subplate

Postcentral gyrus

13

Paracentral lobule

2

1

Cingulate gyrus

Indusium griseum

White matter

3

Cingulum

Lateral ventricle

Corpus callosum

4

Fornix

Superior temporal gyrus

10

11

Thalamus

Caudate (tail)
Stria terminalis

Pulvinar

Fimbria/fornix

9

Middle temporal gyrus

Entorhinal cortex

Dentate gyrus

6

Hippo-campus

Tectum

Superior colliculus

Subiculum

12

Cerebral aqueduct

Lateral ventricle
Ammon's horn

8

5

Midbrain

Central gray

7

Inferior temporal gyrus

Temporal lobe

Tegmentum

Fusiform gyrus

Parahippo-campal gyrus

Superior cerebellar peduncle

Raphe nuclear complex

Pons

Nerve V

Pontine gray
Transpontine corticofugal tract

Reticular tegmental nucleus

Middle cerebellar peduncle

PLATE 64A
CR 350 mm
GW 37
Y217-65
Coronal
Section 1181

✳ See this area of cortex in
Plates 75 and 76.

See detail of the
brain core in
Plates 91A and B.

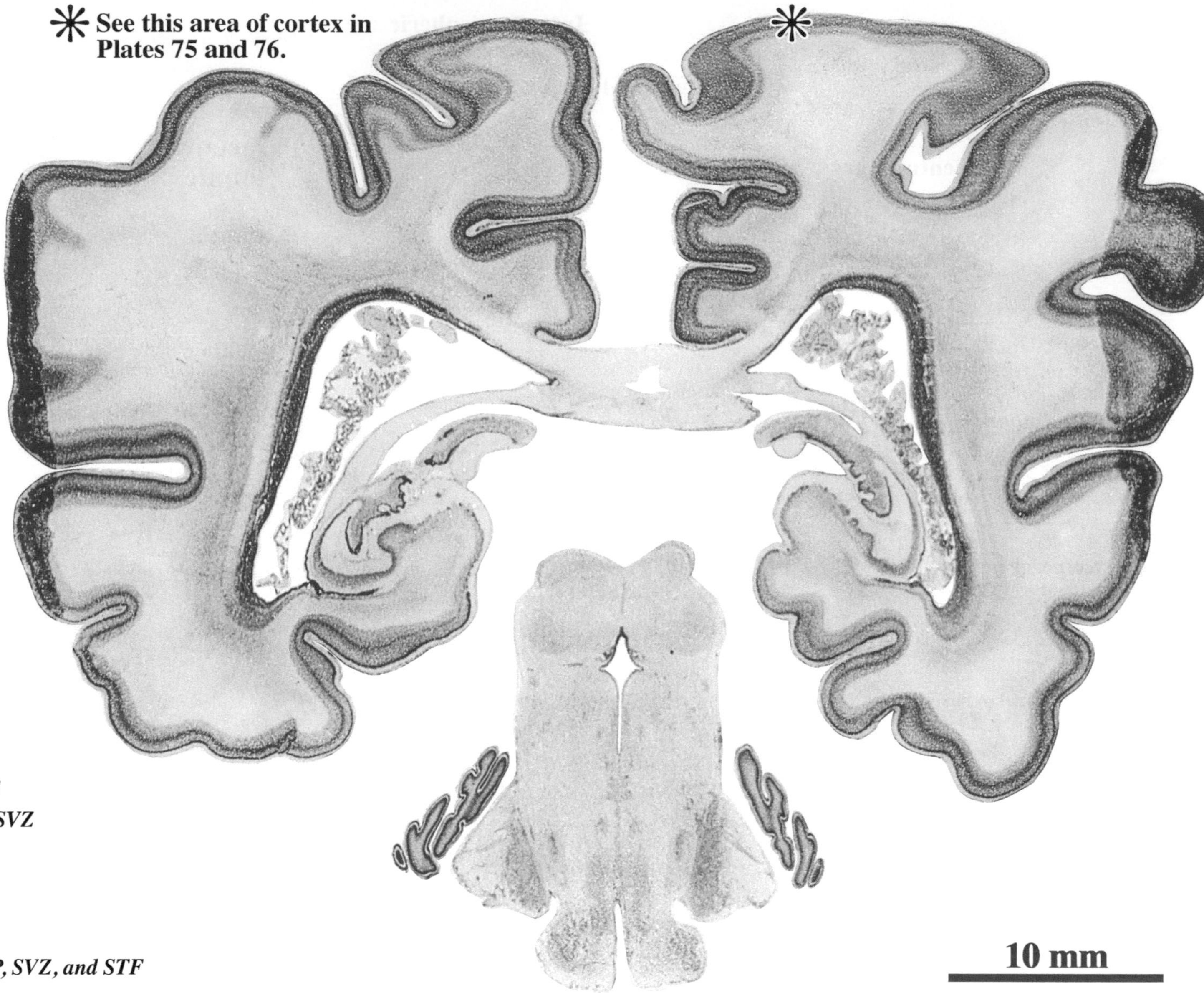

Remnants of the
germinal matrix,
migratory streams,
and transitional fields

1 *Paracentral NEP and SVZ*

2 *Paracentral STF*

3 *Callosal GEP*

4 *Fornical GEP*

5 *Parahippocampal NEP, SVZ, and STF*

6 *Temporal NEP and SVZ*

7 *Temporal STF*

8 *Alvear GEP*

9 *Subgranular zone (dentate)*

10 *Mesencephalic G/EP*

11 *External germinal layer (cerebellum)*

12 *Subpial granular layer (cortical)*

GEP - Glioepithelium
G/EP - Glioepithelium/ependyma
NEP - Neuroepithelium
STF - Stratified transitional field
SVZ - Subventricular zone

10 mm

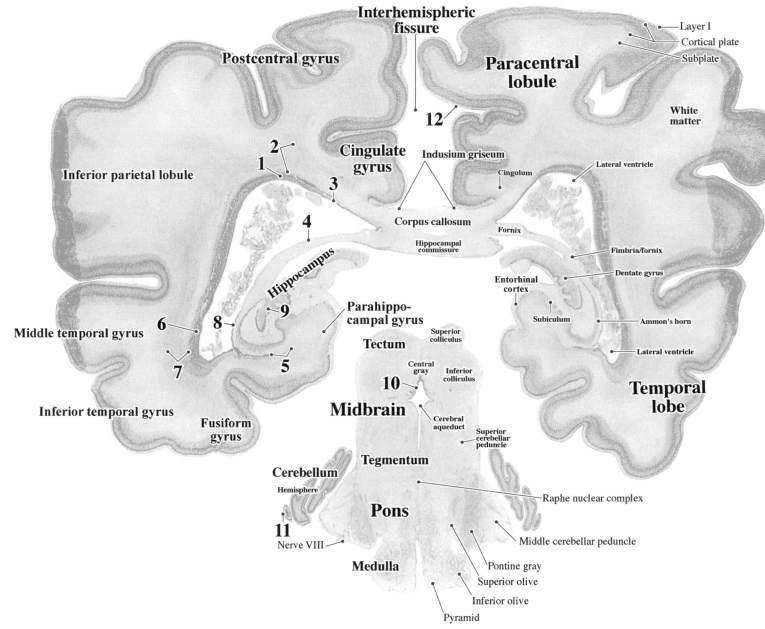

Interhemispheric fissure

Postcentral gyrus

Paracentral lobule

Layer I
Cortical plate
Subplate

White matter

Cingulate gyrus

Indusium griseum

Cingulum

Lateral ventricle

2
1
3
4

Inferior parietal lobule

Corpus callosum

Fornix

Hippocampal commissure

Fimbria/fornix

Dentate gyrus

Hippocampus

Entorhinal cortex

9

Parahippo-campal gyrus

Subiculum

Ammon's horn

6
8
7
5

Middle temporal gyrus

Superior colliculus

Tectum

Lateral ventricle

Temporal lobe

Central gray
Inferior colliculus

Inferior temporal gyrus

Fusiform gyrus

10

Midbrain

Cerebral aqueduct

Superior cerebellar peduncle

Tegmentum

Cerebellum

Hemisphere

Raphe nuclear complex

Pons

11

Nerve VIII

Middle cerebellar peduncle

Medulla

Pontine gray

Superior olive

Inferior olive

Pyramid

138

PLATE 65A
CR 350 mm
GW 37
Y217-65
Coronal
Section 1221

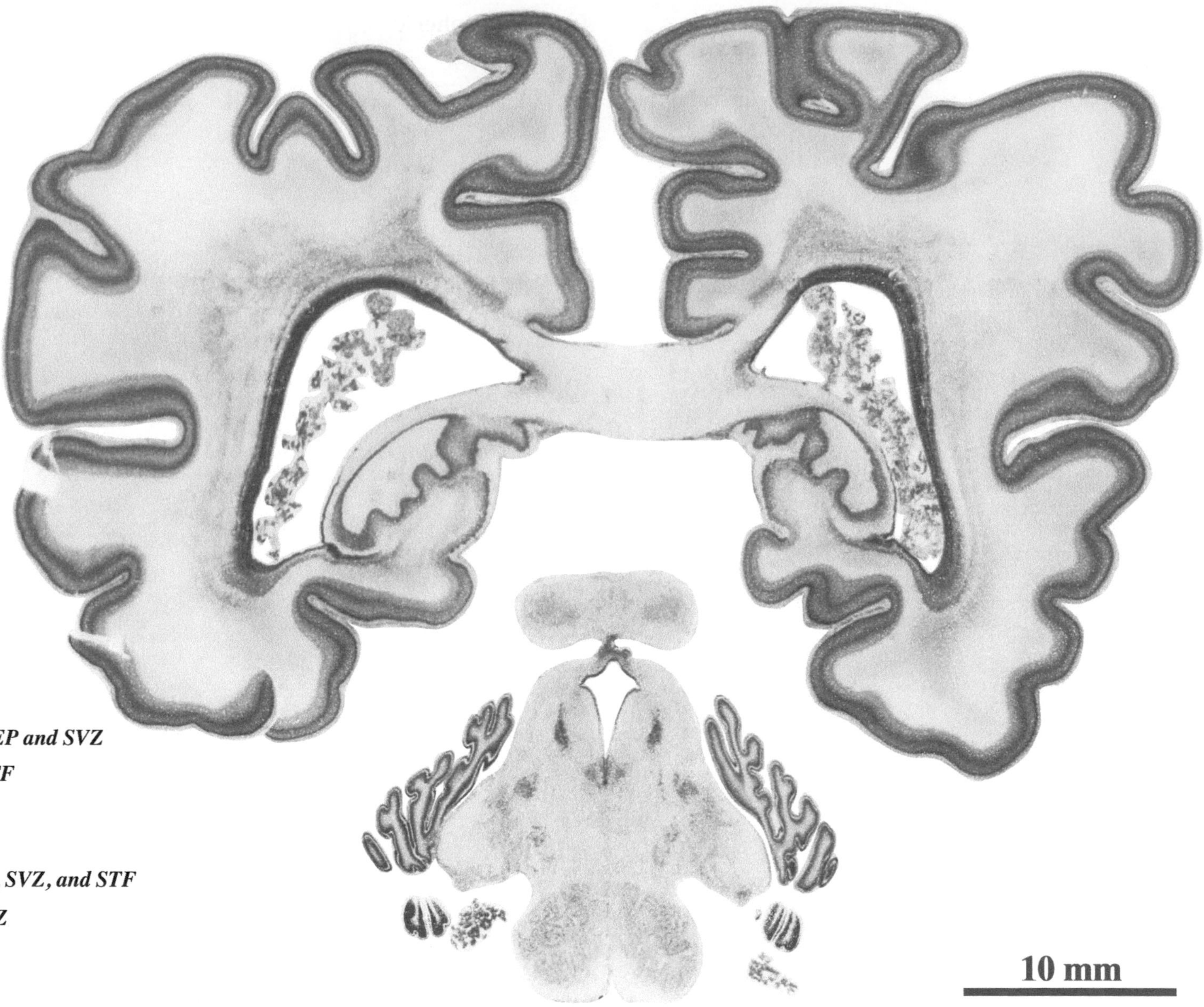

See detail of the
brain core in
Plates 92A and B.

Remnants of the
germinal matrix,
migratory streams,
and transitional fields

1 *Paracentral/parietal NEP and SVZ*

2 *Paracentral/parietal STF*

3 *Callosal GEP*

4 *Fornical GEP*

5 *Parahippocampal NEP, SVZ, and STF*

6 *Temporal NEP and SVZ*

7 *Temporal STF*

8 *Alvear GEP*

9 *Mesencephalic G/EP*

10 *External germinal layer (cerebellum)*

11 *Subpial granular layer (cortical)*

GEP - Glioepithelium
G/EP - Glioepithelium/ependyma
NEP - Neuroepithelium
STF - Stratified transitional field
SVZ - Subventricular zone

10 mm

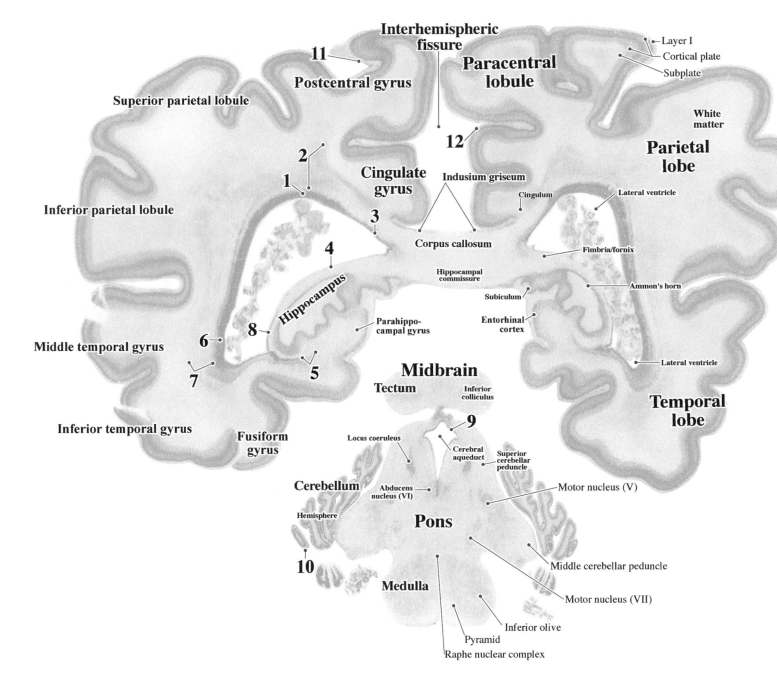

Interhemispheric fissure

Paracentral lobule

Layer I
Cortical plate
Subplate

11

Postcentral gyrus

Superior parietal lobule

White matter

Parietal lobe

2

1

Cingulate gyrus

Indusium griseum

Cingulum

Lateral ventricle

12

Inferior parietal lobule

3

Corpus callosum

Fimbria/fornix

4

Hippocampal commissure

Ammon's horn

Subiculum

Hippocampus

Entorhinal cortex

6

8

Parahippo-campal gyrus

Middle temporal gyrus

Lateral ventricle

7

5

Midbrain

Tectum

Inferior colliculus

Temporal lobe

Inferior temporal gyrus

9

Fusiform gyrus

Locus coeruleus

Cerebral aqueduct

Superior cerebellar peduncle

Cerebellum

Abducens nucleus (VI)

Motor nucleus (V)

Hemisphere

Pons

Middle cerebellar peduncle

10

Motor nucleus (VII)

Medulla

Inferior olive

Pyramid

Raphe nuclear complex

PLATE 66A
CR 350 mm
GW 37
Y217-65
Coronal
Section 1311

See detail of the
brain core in
Plates 93A and B.

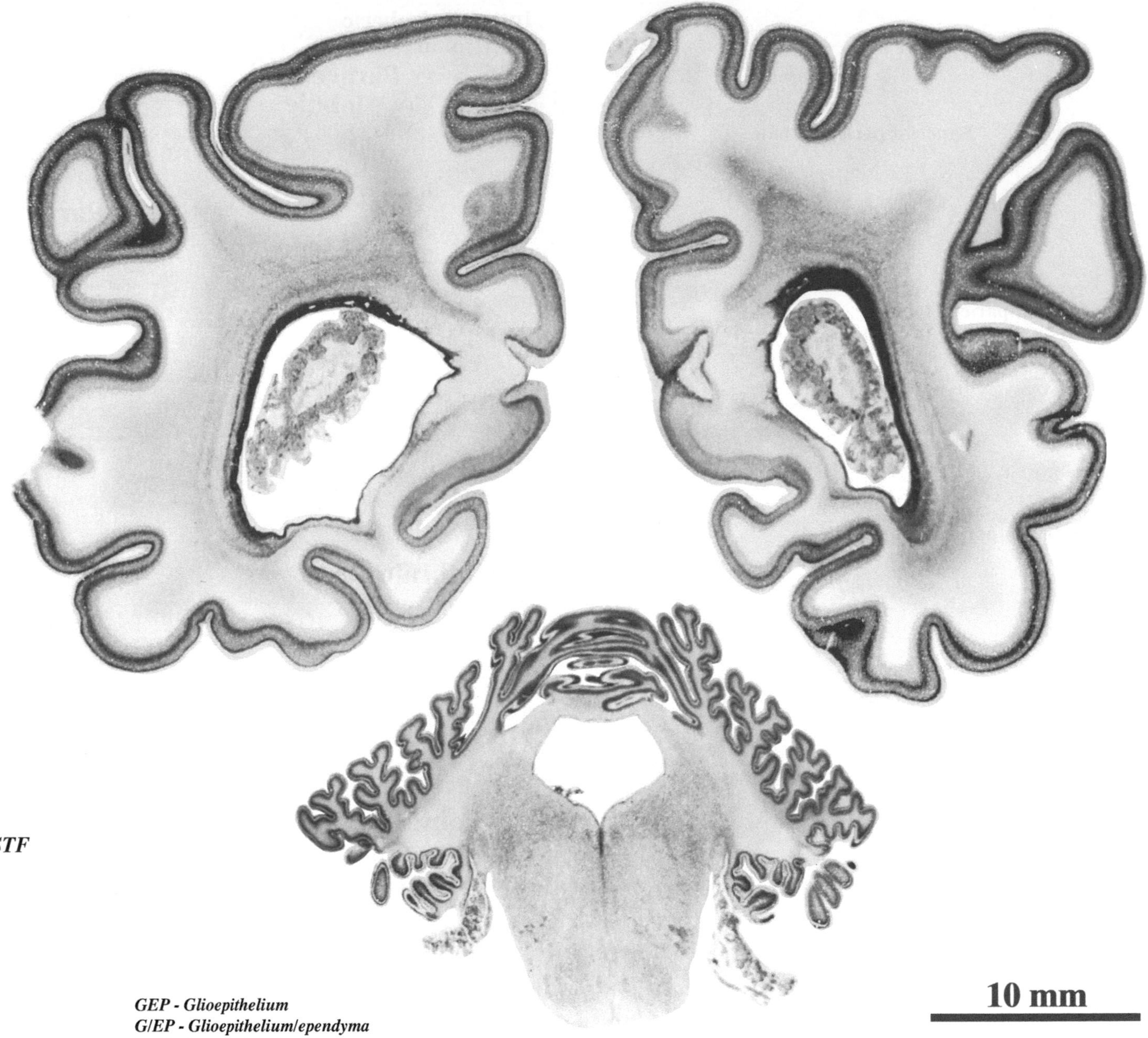

Remnants of the
germinal matrix,
migratory streams,
and transitional fields

1 *Parietal NEP and SVZ*
2 *Parietal STF*
3 *Callosal GEP*
4 *Cingulate NEP, SVZ, and STF*
5 *Occipital NEP and SVZ*
6 *Occipital STF*
7 *Temporal NEP and SVZ*
8 *Temporal STF*
9 *Medullary G/EP*
10 *Pontine G/EP*
11 *External germinal layer (cerebellum)*
12 *Subpial granular layer (cortical)*

GEP - Glioepithelium
G/EP - Glioepithelium/ependyma
NEP - Neuroepithelium
STF - Stratified transitional field
SVZ - Subventricular zone

10 mm

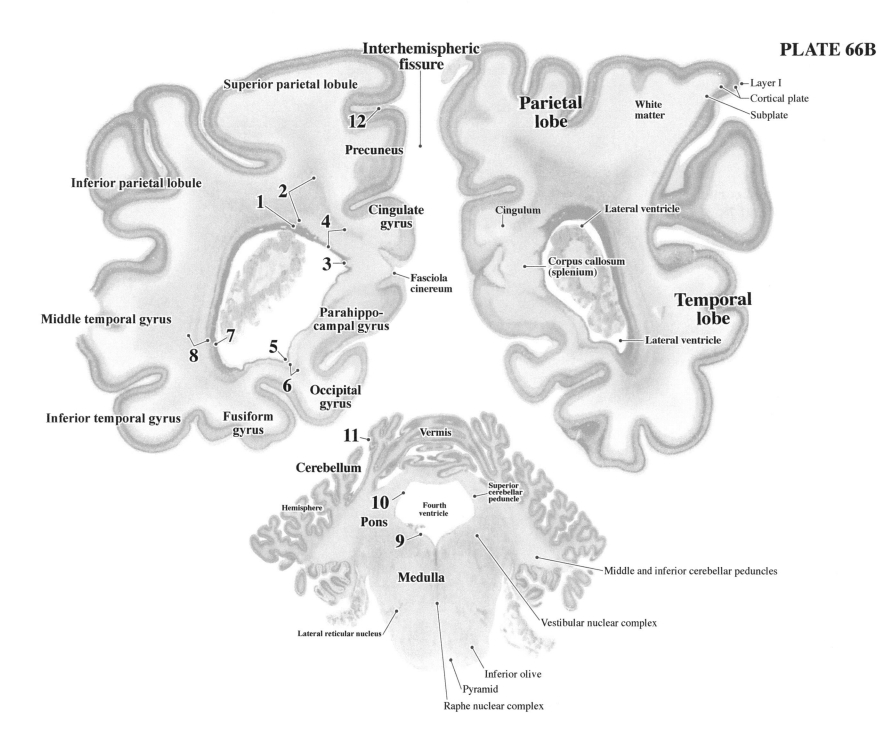

Interhemispheric fissure

Superior parietal lobule

Parietal lobe

White matter

Layer I

Cortical plate

Subplate

12

Precuneus

Inferior parietal lobule

2

1

Cingulum

Lateral ventricle

4

Cingulate gyrus

3

Corpus callosum (splenium)

Fasciola cinereum

Temporal lobe

Middle temporal gyrus

Parahippo-campal gyrus

Lateral ventricle

7

5

8

6

Occipital gyrus

Inferior temporal gyrus

Fusiform gyrus

Vermis

11

Cerebellum

Superior cerebellar peduncle

Hemisphere

10

Fourth ventricle

Pons

9

Middle and inferior cerebellar peduncles

Medulla

Vestibular nuclear complex

Lateral reticular nucleus

Inferior olive

Pyramid

Raphe nuclear complex

PLATE 67A
CR 350 mm
GW 37
Y217-65
Coronal
Section 1371

✳ **See this area of cortex**
   in Plates 77 and 78

See detail of the
brain core and
cerebellum in
Plates 94A and B.

Remnants of the
germinal matrix,
migratory streams,
and transitional fields

1 *Parietal NEP and SVZ*

2 *Parietal STF*

3 *Cingulate NEP and SVZ*

4 *Cingulate STF*

5 *Occipital NEP and SVZ*

6 *Occipital STF*

7 *Temporal NEP and SVZ*

8 *Temporal STF*

9 *External germinal layer (cerebellum)*

10 *Germinal trigone (cerebellum)*

11 *Medullary G/EP*

12 *Subpial granular layer (cortical)*

G/EP - Glioepithelium/ependyma
NEP - Neuroepithelium
STF - Stratified transitional field
SVZ - Subventricular zone

10 mm

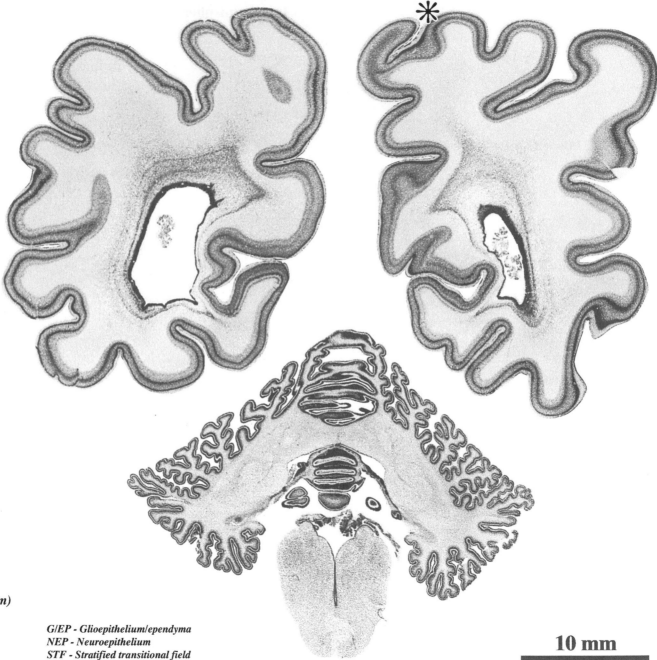

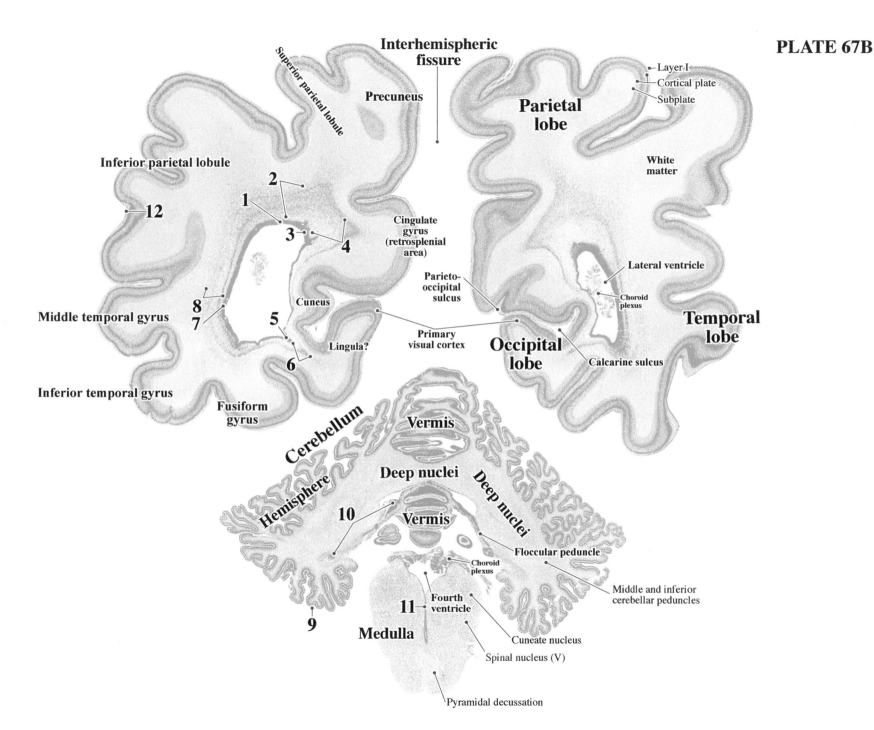

Interhemispheric fissure

Precuneus

Superior parietal lobule

Layer I
Cortical plate
Subplate

Parietal lobe

White matter

Inferior parietal lobule

2
1
12
3
4

Cingulate gyrus (retrosplenial area)

Lateral ventricle

Choroid plexus

Parieto-occipital sulcus

Middle temporal gyrus

8
7

Cuneus

5

Primary visual cortex

Occipital lobe

Temporal lobe

Lingula?

Calcarine sulcus

6

Inferior temporal gyrus

Fusiform gyrus

Cerebellum

Vermis

Deep nuclei

Deep nuclei

Hemisphere

10

Vermis

Floccular peduncle

9

Choroid plexus

Middle and inferior cerebellar peduncles

11

Fourth ventricle

Medulla

Cuneate nucleus

Spinal nucleus (V)

Pyramidal decussation

PLATE 68A
CR 350 mm
GW 37
Y217-65
Coronal
Section 1501

See detail of the
cerebellum in
Plates 95A and B.

Remnants of the
germinal matrix,
migratory streams,
and transitional fields

1 *Parietal NEP and SVZ*

2 *Parietal STF*

3 *Occipital NEP and SVZ*

4 *Occipital STF*

5 *Temporal NEP and SVZ*

6 *Temporal STF*

7 *External germinal layer (cerebellum)*

8 *Subpial granular layer (cortical)*

NEP - Neuroepithelium
STF - Stratified transitional field
SVZ - Subventricular zone

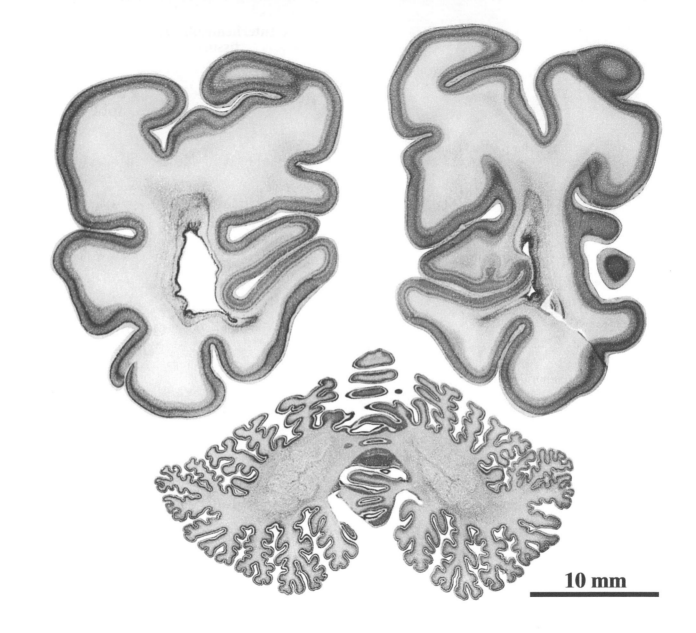

10 mm

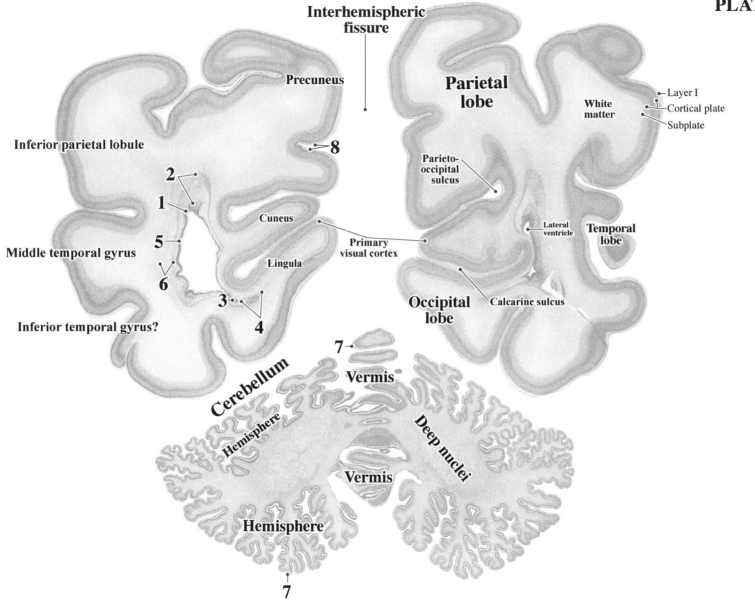

Interhemispheric fissure

Precuneus

Parietal lobe

Layer I
Cortical plate
White matter
Subplate

Inferior parietal lobule

8

Parieto-occipital sulcus

2

1

Cuneus

Primary visual cortex

Lateral ventricle

Temporal lobe

5

Middle temporal gyrus

Lingula

6

3

4

Calcarine sulcus

Occipital lobe

Inferior temporal gyrus?

7

Cerebellum

Vermis

Hemisphere

Deep nuclei

Vermis

Hemisphere

7

PLATE 69A
CR 350 mm
GW 37
Y217-65
Coronal
Section 1611

See detail of the
cerebellum in
Plates 95A and B.

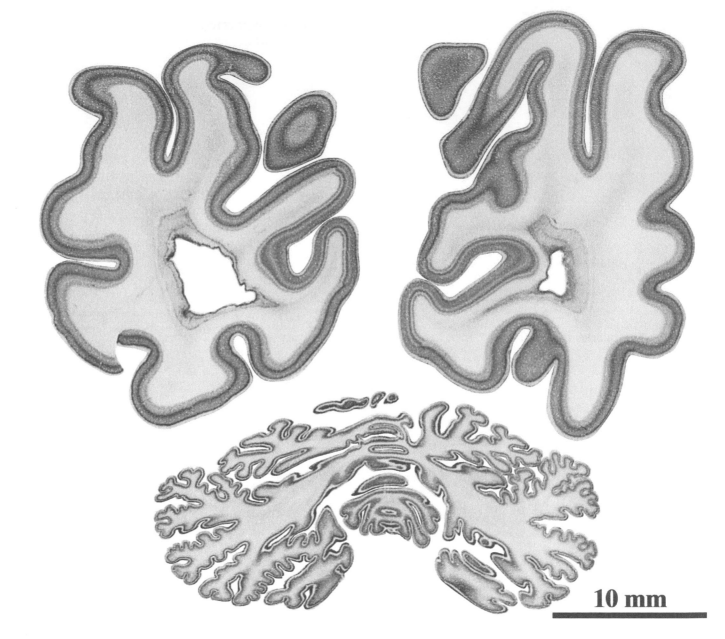

Remnants of the
germinal matrix,
migratory streams,
and transitional fields

1 *Parietal NEP and SVZ*

2 *Parietal STF*

3 *Occipital NEP and SVZ*

4 *Occipital STF*

5 *Temporal NEP and SVZ*

6 *Temporal STF*

7 *External germinal layer (cerebellum)*

8 *Subpial granular layer (cortical)*

NEP - Neuroepithelium
STF - Stratified transitional field
SVZ - Subventricular zone

10 mm

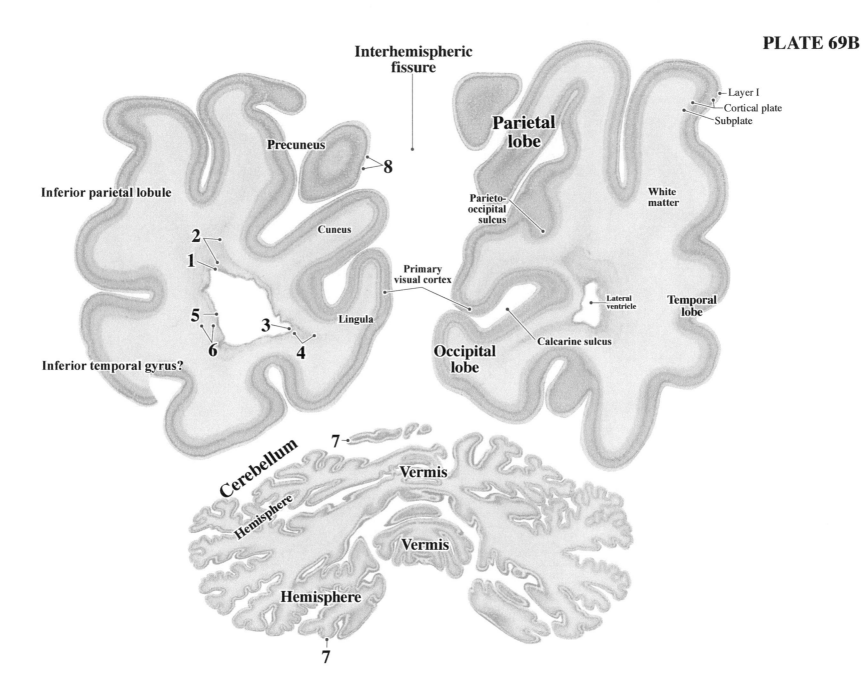

Interhemispheric
fissure

Precuneus

Inferior parietal lobule

Cuneus

2

1

Lingula

5

3

6

4

Inferior temporal gyrus?

Parietal
lobe

Layer I
Cortical plate
Subplate

Parieto-
occipital
sulcus

White
matter

Primary
visual cortex

Lateral
ventricle

Temporal
lobe

Occipital
lobe

Calcarine sulcus

Cerebellum

7

Vermis

Hemisphere

Vermis

Hemisphere

7

PLATE 70A
CR 350 mm
GW 37
Y217-65
Coronal
Section 1711

See detail of the
cerebellum in
Plates 97A and B.

✳ **See this area of cortex**
  in Plates 79 and 80

Remnants of the
germinal matrix,
migratory streams,
and transitional fields

1 *Parietal NEP and SVZ*

2 *Parietal STF*

3 *Occipital NEP and SVZ*

4 *Occipital STF*

5 *External germinal layer (cerebellum)*

6 *Subpial granular layer (cortical)*

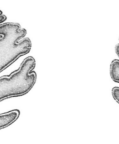

NEP - Neuroepithelium
STF - Stratified transitional field
SVZ - Subventricular zone

10 mm

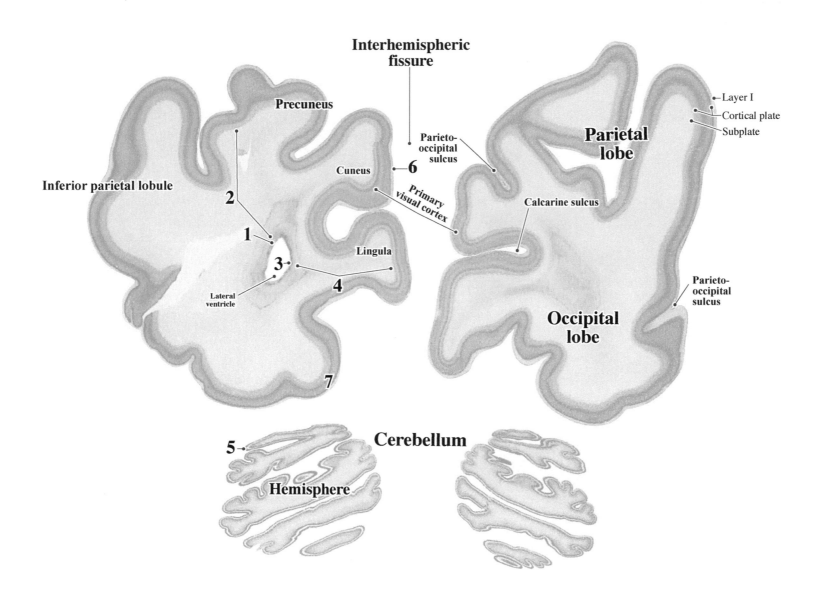

Interhemispheric fissure

Precuneus

Inferior parietal lobule

Parieto-occipital sulcus

Cuneus

6

Parietal lobe

Layer I

Cortical plate

Subplate

2

Primary visual cortex

Calcarine sulcus

1

Lingula

3

4

Lateral ventricle

Parieto-occipital sulcus

Occipital lobe

7

Cerebellum

5

Hemisphere

PLATE 71
CR 350 mm, GW 37, Y217-65
Coronal, Section 761
FRONTAL CORTEX

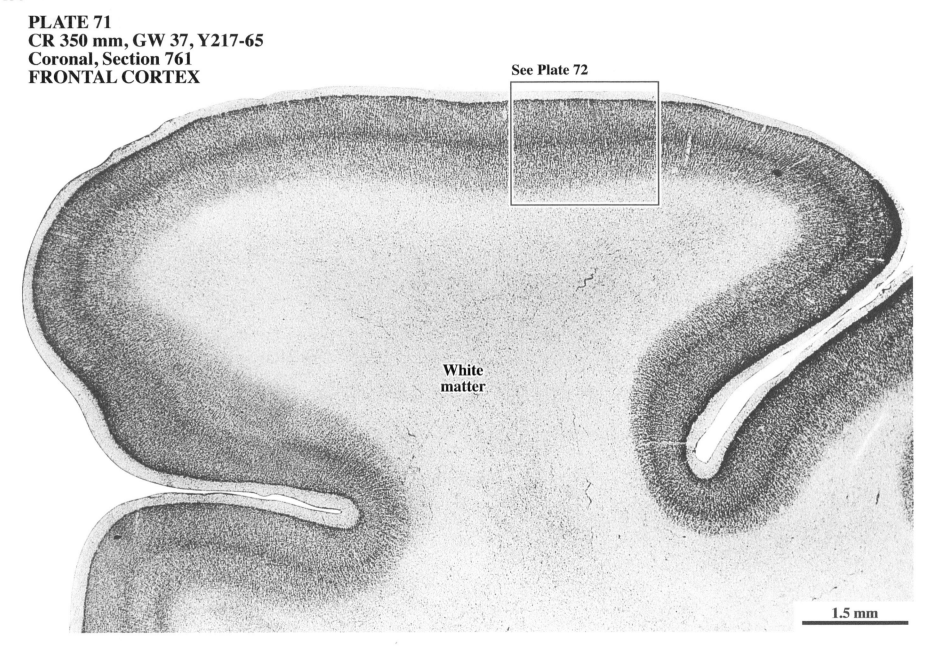

See Plate 72

White
matter

1.5 mm

See the entire Section 761 in Plate 57.

PLATE 72
CR 350 mm, GW 37, Y217-65, Coronal, Section 761

FRONTAL CORTEX

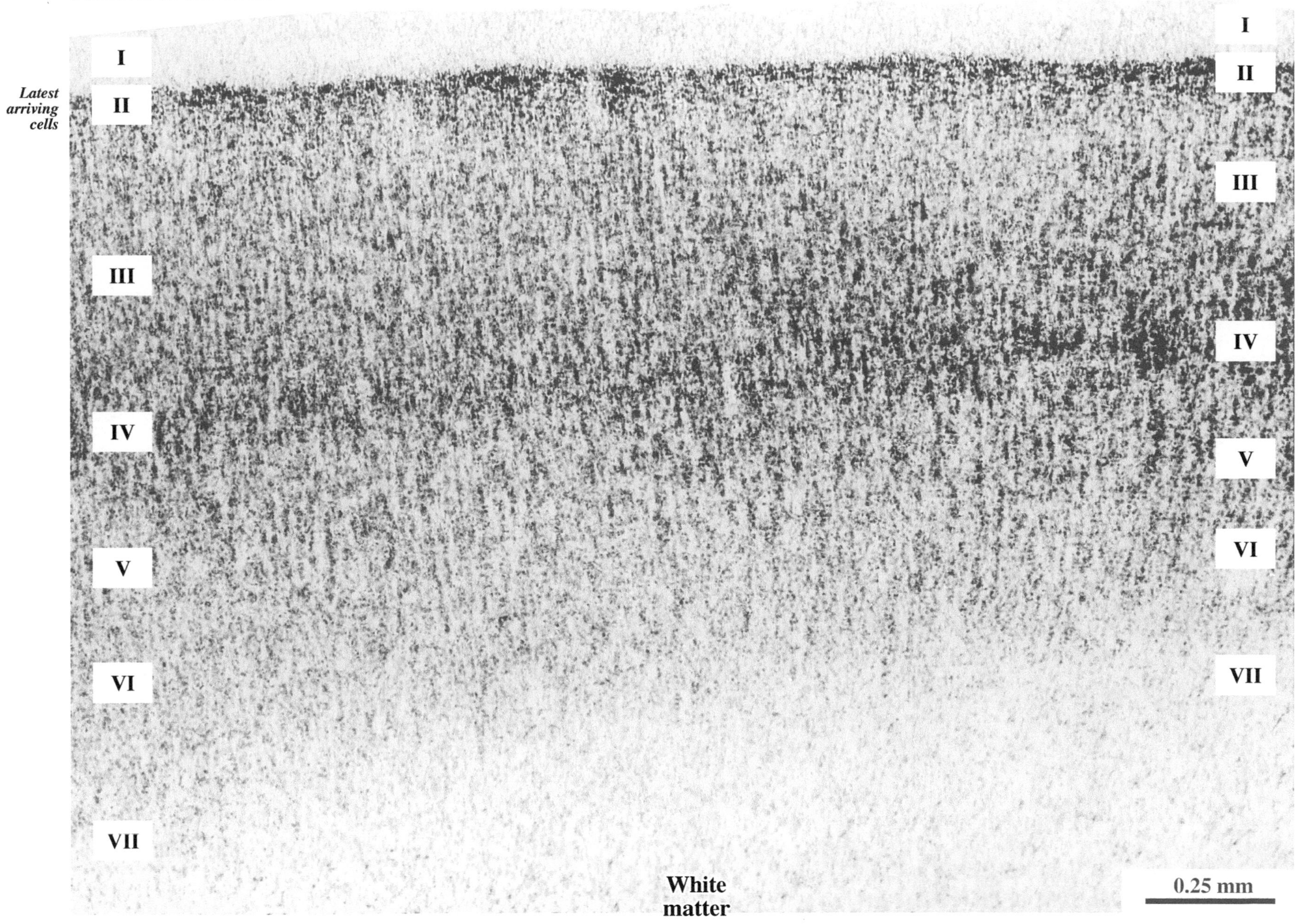

I

*Latest
arriving
cells*

II

III

IV

V

VI

VII

I

II

III

IV

V

VI

VII

**White
matter**

0.25 mm

152

PLATE 73
CR 350 mm, GW 37, Y217-65
Coronal, Section 1081
CORTEX OF THE
PRECENTRAL GYRUS
(Primary motor cortex)

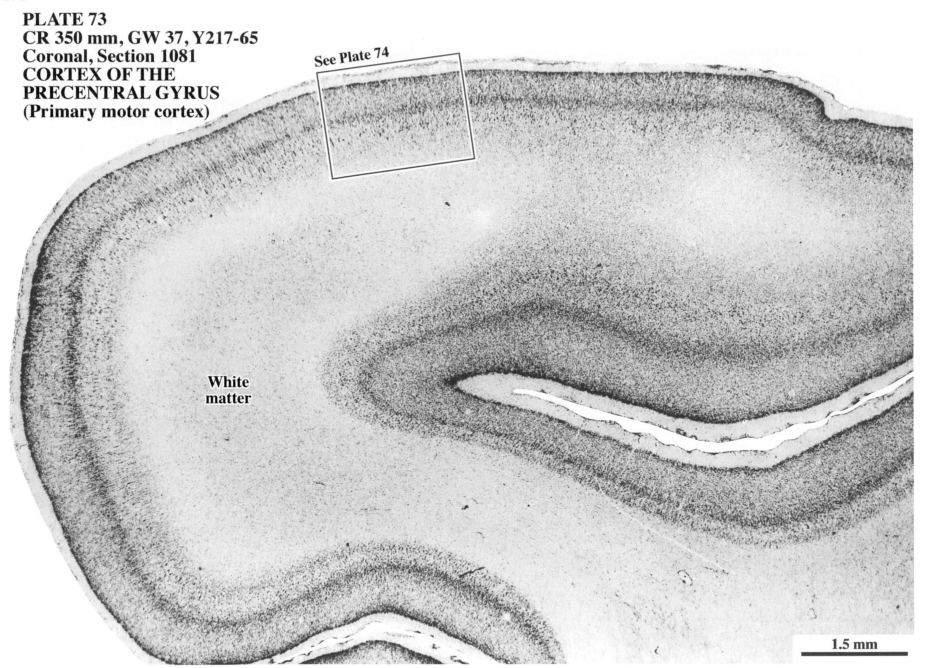

See Plate 74

White
matter

1.5 mm

See the entire Section 1081 in Plate 62.

PLATE 74
CR 350 mm, GW 37, Y217-65, Coronal, Section 1081

CORTEX OF THE PRECENTRAL GYRUS

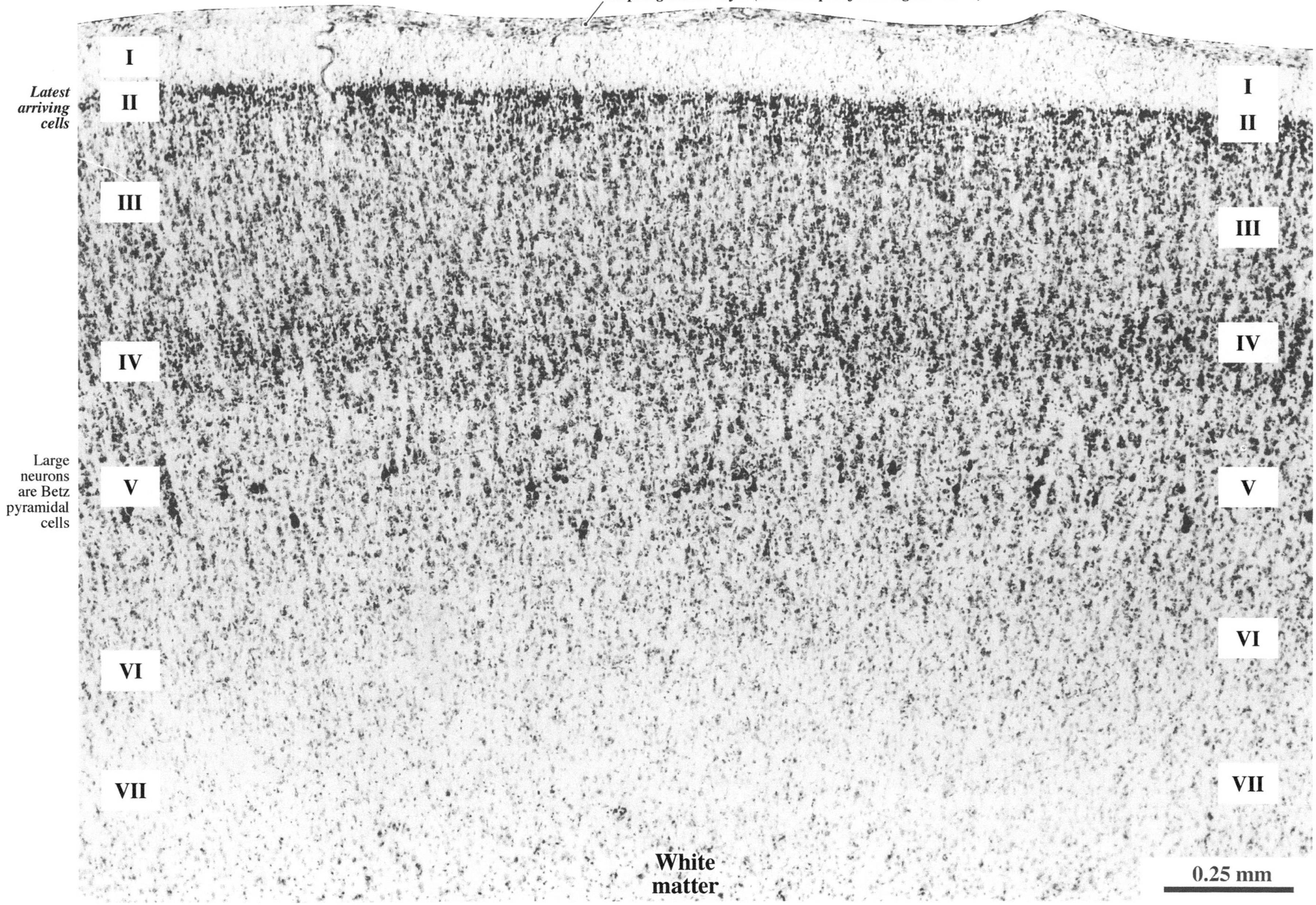

Subpial granular layer (transient proliferative glial matrix)

I

*Latest
arriving
cells*

II

III

IV

Large
neurons
are Betz
pyramidal
cells

V

VI

VII

I

II

III

IV

V

VI

VII

**White
matter**

0.25 mm

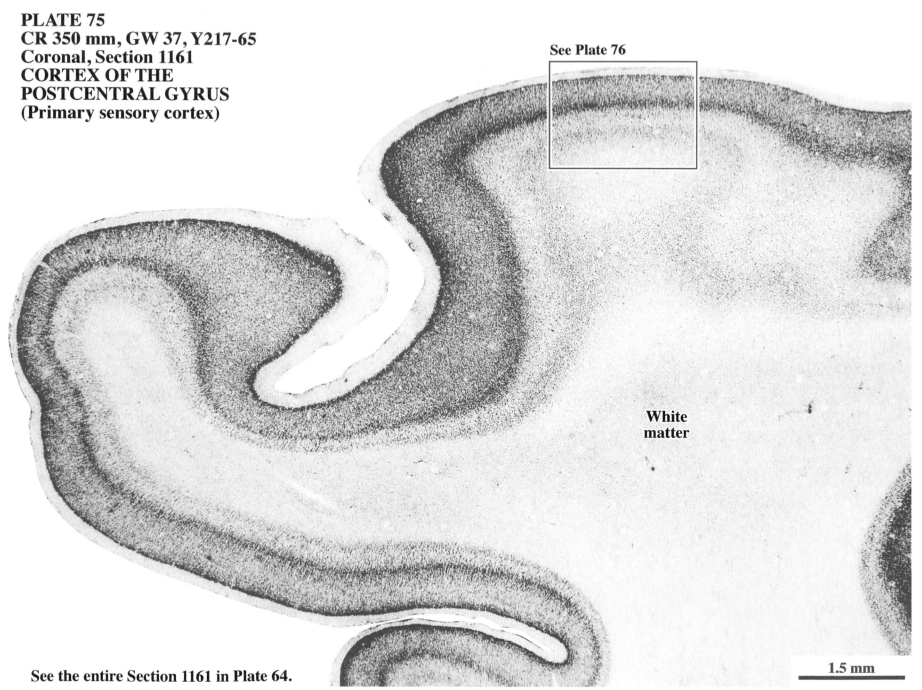

154

PLATE 75
CR 350 mm, GW 37, Y217-65
Coronal, Section 1161
CORTEX OF THE
POSTCENTRAL GYRUS
(Primary sensory cortex)

See Plate 76

White matter

1.5 mm

See the entire Section 1161 in Plate 64.

PLATE 76
CR 350 mm, GW 37, Y217-65, Coronal, Section 1161

CORTEX OF THE POSTCENTRAL GYRUS

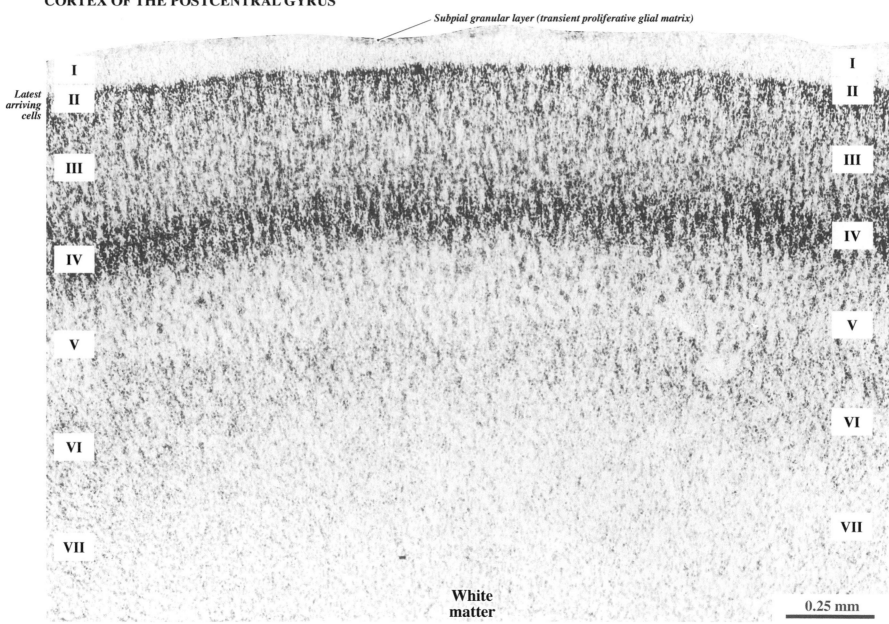

Subpial granular layer (transient proliferative glial matrix)

I

*Latest
arriving
cells*

II

III

IV

V

VI

VII

I

II

III

IV

V

VI

VII

**White
matter**

0.25 mm

156

PLATE 77
CR 350 mm, GW 37, Y217-65
Coronal, Section 1371
PARIETAL CORTEX

See Plate 78

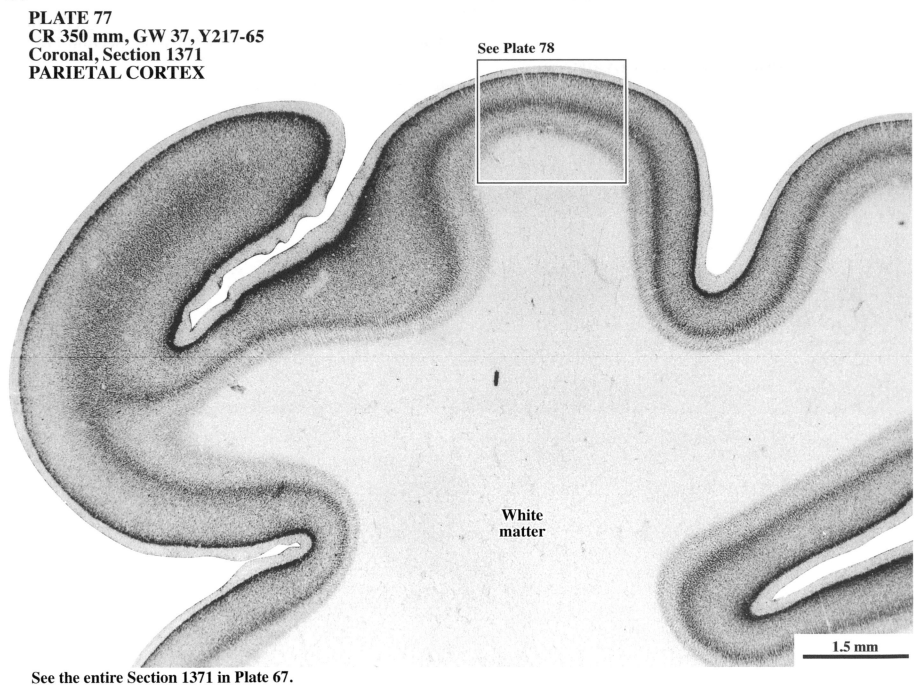

White
matter

1.5 mm

See the entire Section 1371 in Plate 67.

PLATE 78
CR 350 mm, GW 37, Y217-65, Coronal, Section 1371

PARIETAL CORTEX

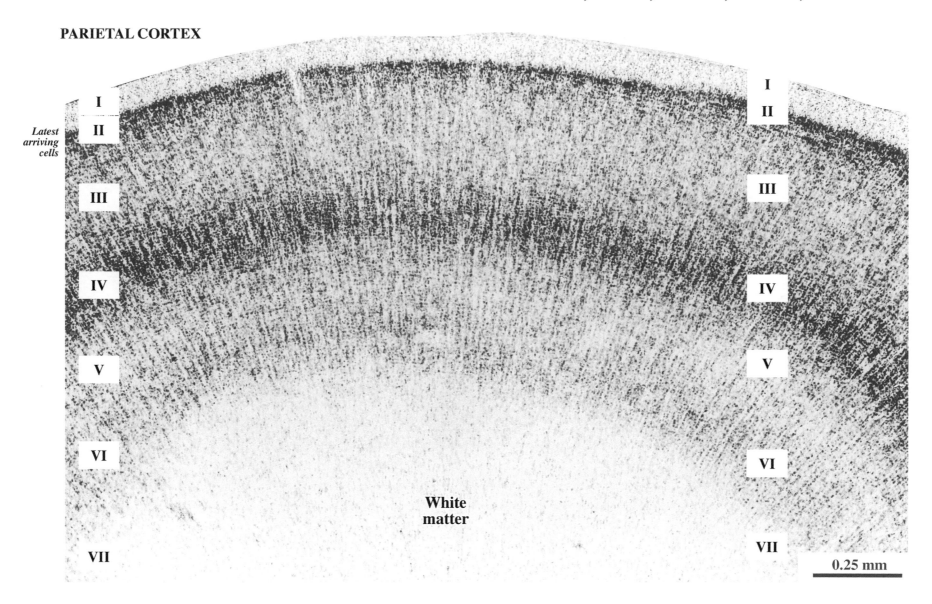

Latest arriving cells

I

II

III

IV

V

VI

VII

I
II

III

IV

V

VI

VII

White matter

0.25 mm

PLATE 79
CR 350 mm, GW 37, Y217-65
Coronal, Section 1751
STRIATE/PERISTRIATE
CORTICAL
TRANSITION
AREA

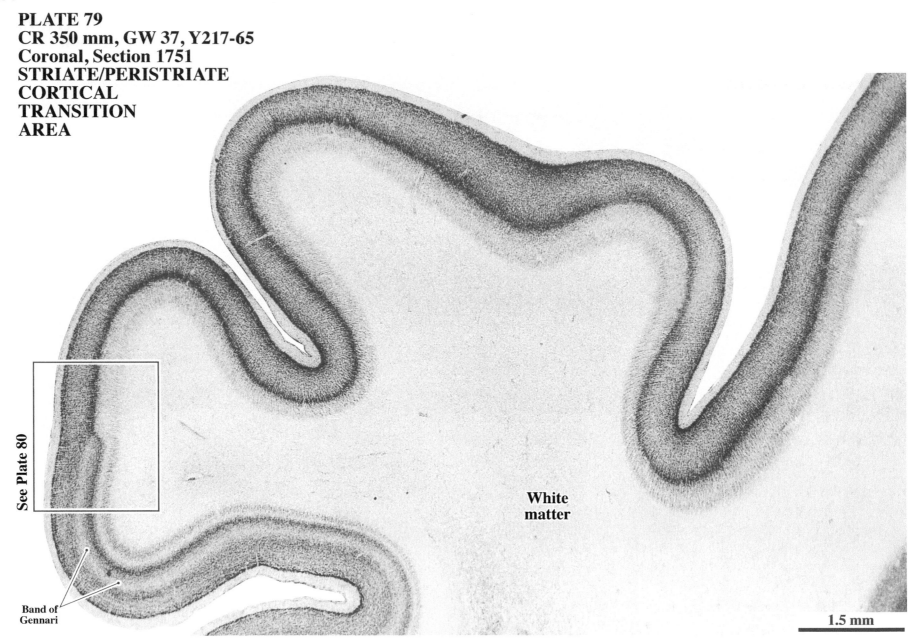

See Plate 80

White
matter

Band of
Gennari

1.5 mm

See the nearby entire Section 1711 in Plate 70.

PLATE 80
CR 350 mm, GW 37, Y217-65, Coronal, Section 1751

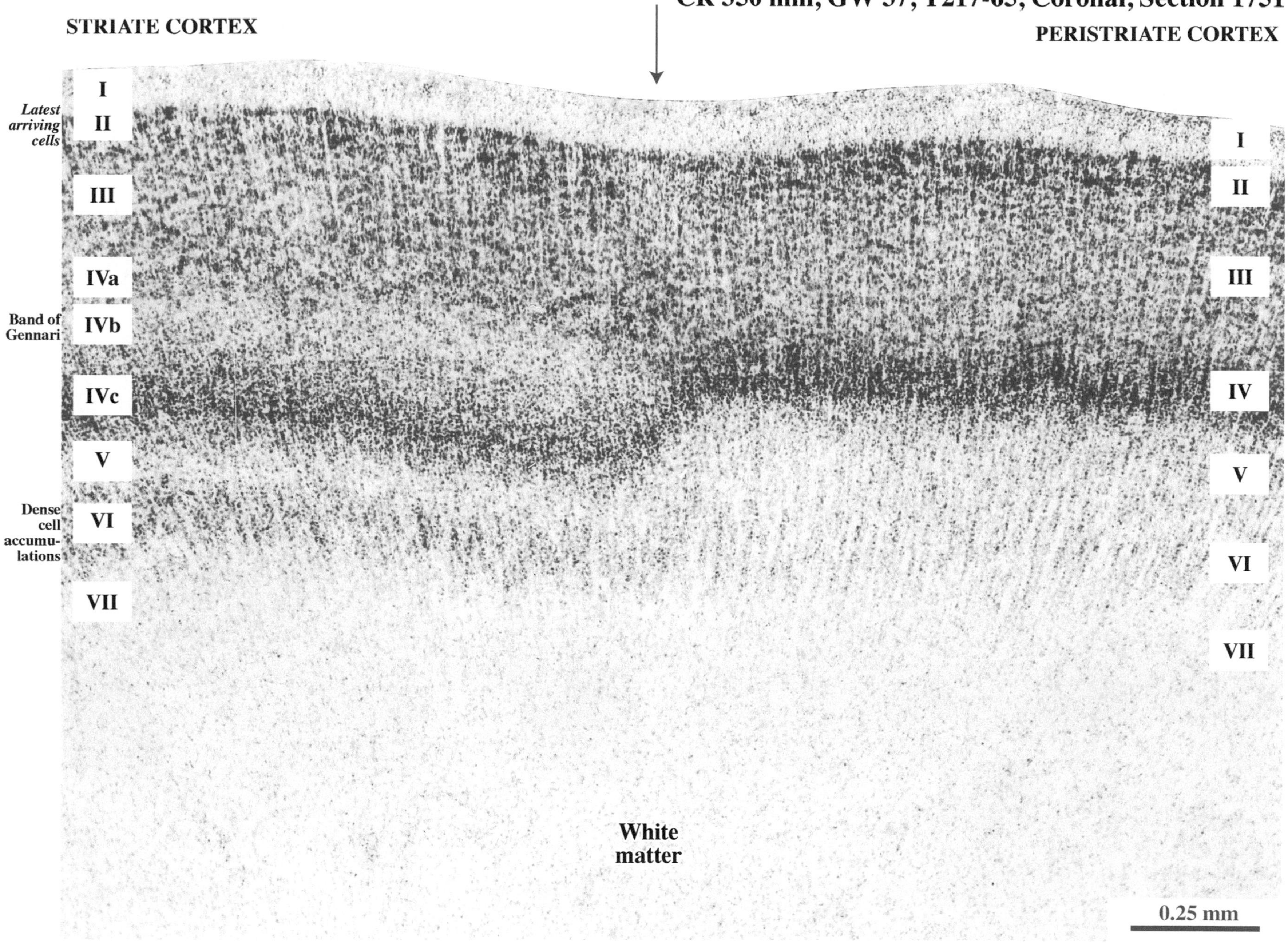

STRIATE CORTEX

PERISTRIATE CORTEX

I
II

Latest arriving cells

III

IVa

Band of Gennari

IVb

IVc

V

Dense cell accumulations

VI

VII

I

II

III

IV

V

VI

VII

White matter

0.25 mm

PLATE 81A
CR 350 mm
GW 37
Y217-65
Coronal
Section 611

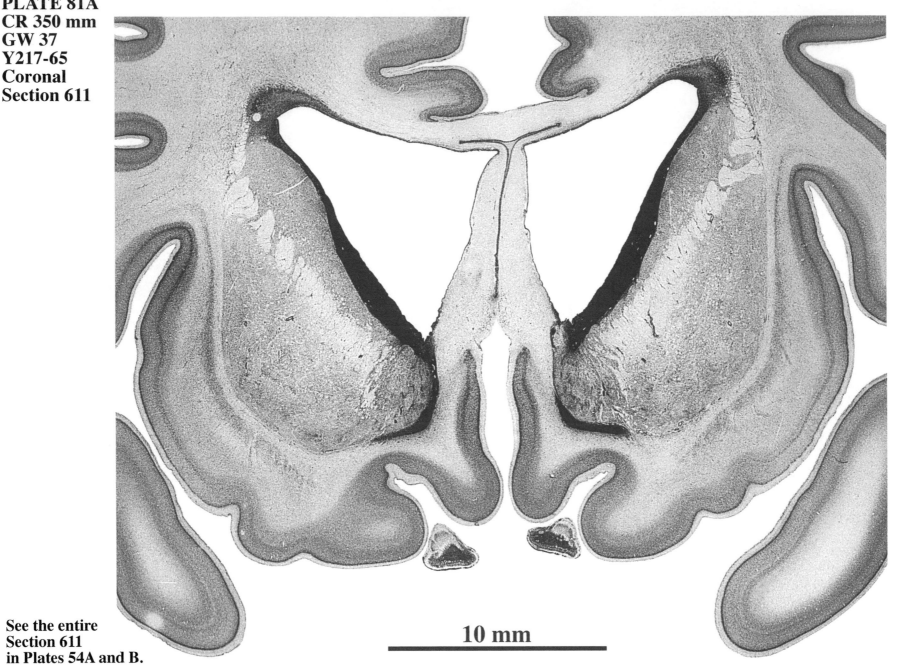

See the entire
Section 611
in Plates 54A and B.

10 mm

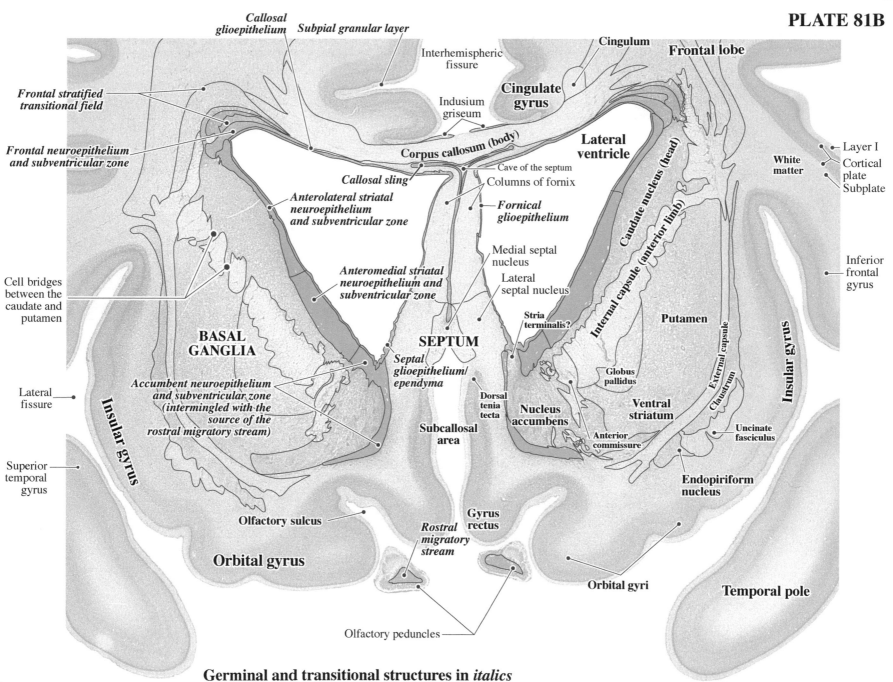

Callosal glioepithelium   *Subpial granular layer*

Interhemispheric fissure

Cingulum   **Frontal lobe**

Frontal stratified transitional field

Indusium griseum

Cingulate gyrus

Lateral ventricle

Frontal neuroepithelium and subventricular zone

Corpus callosum (body)

Caudate nucleus (head)

White matter

Layer I
Cortical plate
Subplate

Callosal sling

Cave of the septum

Columns of fornix

Anterolateral striatal neuroepithelium and subventricular zone

Fornical glioepithelium

Internal capsule (anterior limb)

Inferior frontal gyrus

Anteromedial striatal neuroepithelium and subventricular zone

Medial septal nucleus

Lateral septal nucleus

Cell bridges between the caudate and putamen

Putamen

Stria terminalis?

BASAL GANGLIA

SEPTUM

Insular gyrus

Globus pallidus

Accumbent neuroepithelium and subventricular zone (intermingled with the source of the rostral migratory stream)

Septal glioepithelium/ependyma

External capsule

Claustrum

Lateral fissure

Dorsal tenia tecta

Nucleus accumbens

Ventral striatum

Insular gyrus

Subcallosal area

Uncinate fasciculus

Superior temporal gyrus

Insular gyrus

Anterior commissure

Endopiriform nucleus

Olfactory sulcus

Gyrus rectus

Orbital gyri

Rostral migratory stream

Orbital gyrus

Temporal pole

Olfactory peduncles

Germinal and transitional structures in *italics*

PLATE 82A
CR 350 mm, GW 37, Y217-65, Coronal, Section 691

See the entire Section 691 in Plate 55.

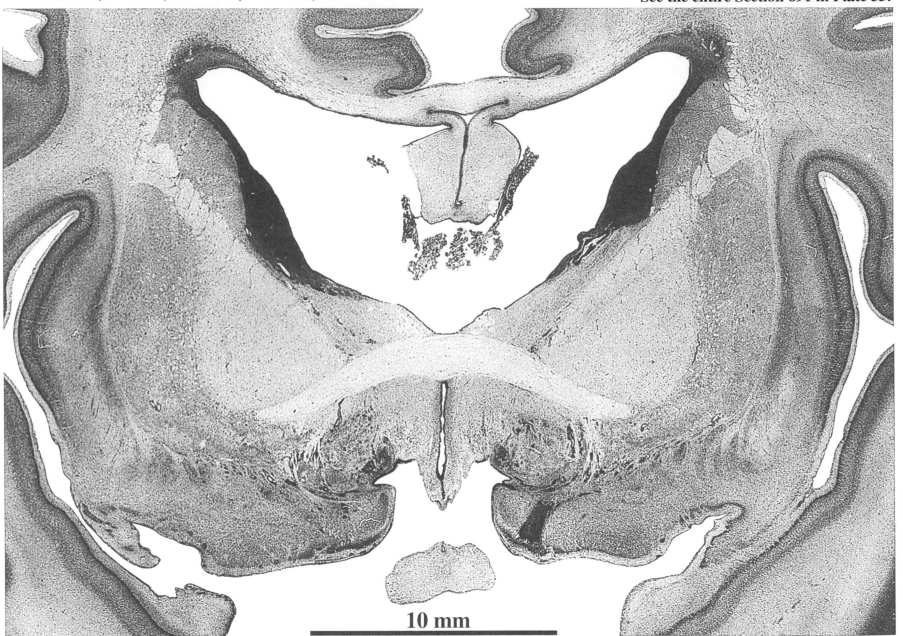

10 mm

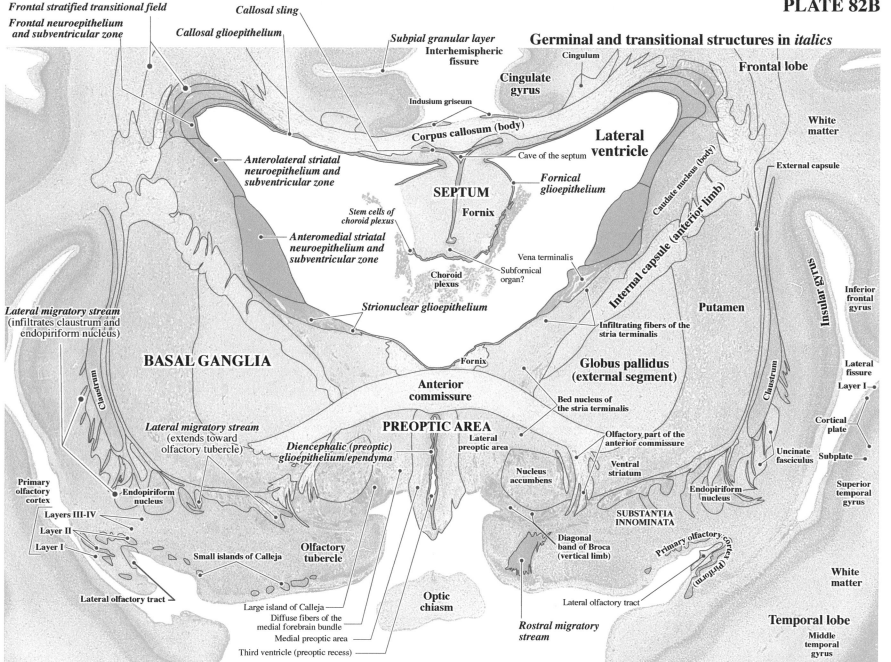

Germinal and transitional structures in *italics*

Frontal stratified transitional field

Frontal neuroepithelium and subventricular zone

Callosal sling

Callosal glioepithelium

Subpial granular layer

Interhemispheric fissure

Cingulum

Cingulate gyrus

Frontal lobe

Indusium griseum

Corpus callosum (body)

Cave of the septum

White matter

Lateral ventricle

Anterolateral striatal neuroepithelium and subventricular zone

Fornical glioepithelium

Caudate nucleus (body)

SEPTUM

Fornix

Stem cells of choroid plexus

Anteromedial striatal neuroepithelium and subventricular zone

Vena terminalis

Subfornical organ?

Choroid plexus

External capsule

Internal capsule (anterior limb)

Strionuclear glioepithelium

Putamen

Infiltrating fibers of the stria terminalis

Insular gyrus

Inferior frontal gyrus

Lateral migratory stream (infiltrates claustrum and endopiriform nucleus)

BASAL GANGLIA

Fornix

Globus pallidus (external segment)

Lateral fissure

Layer I

Claustrum

Anterior commissure

Bed nucleus of the stria terminalis

Cortical plate

Lateral migratory stream (extends toward olfactory tubercle)

PREOPTIC AREA

Lateral preoptic area

Olfactory part of the anterior commissure

Uncinate fasciculus

Subplate

Diencephalic (preoptic) glioepithelium/ependyma

Nucleus accumbens

Ventral striatum

Endopiriform nucleus

Superior temporal gyrus

Primary olfactory cortex

Endopiriform nucleus

SUBSTANTIA INNOMINATA

Layers III-IV

Layer II

Layer I

Small islands of Calleja

Olfactory tubercle

Diagonal band of Broca (vertical limb)

Primary olfactory cortex (piriform)

White matter

Lateral olfactory tract

Large island of Calleja

Diffuse fibers of the medial forebrain bundle

Optic chiasm

Rostral migratory stream

Lateral olfactory tract

Medial preoptic area

Third ventricle (preoptic recess)

Temporal lobe

Middle temporal gyrus

PLATE 83A
CR 350 mm
GW 37
Y217-65
Coronal
Section 721

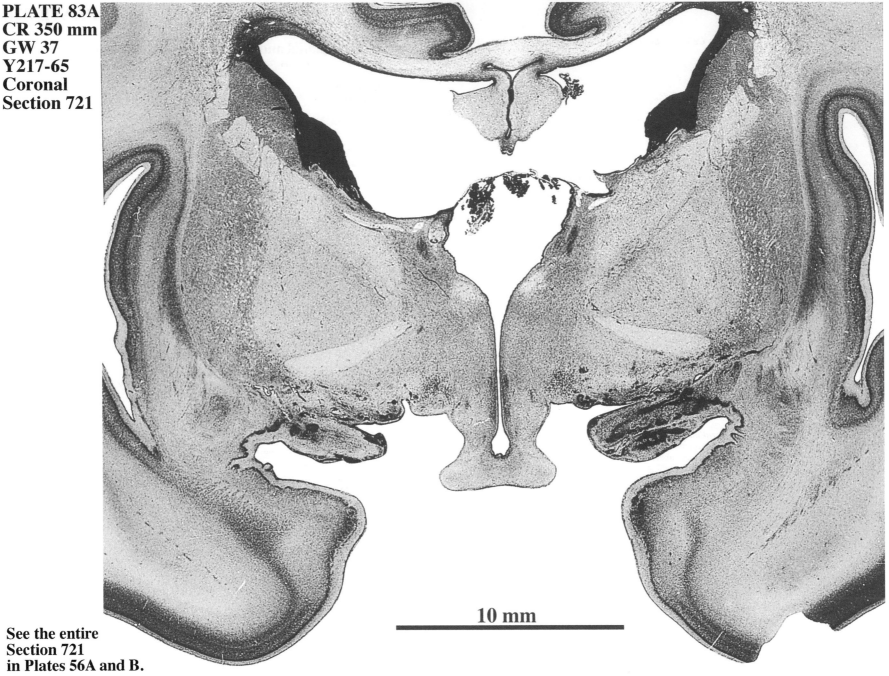

10 mm

**See the entire
Section 721
in Plates 56A and B.**

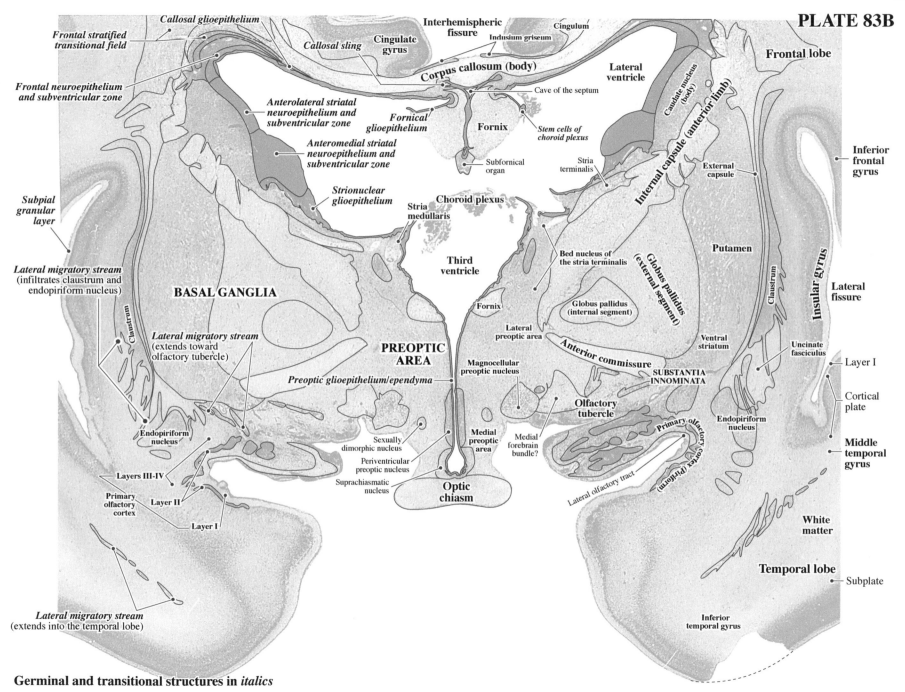

Callosal glioepithelium

Frontal stratified transitional field

Interhemispheric fissure

Cingulum

Indusium griseum

Cingulate gyrus

Callosal sling

Frontal lobe

Lateral ventricle

Frontal neuroepithelium and subventricular zone

Corpus callosum (body)

Cave of the septum

Caudate nucleus (body)

Internal capsule (anterior limb)

Anterolateral striatal neuroepithelium and subventricular zone

Fornical glioepithelium

Fornix

Stem cells of choroid plexus

Anteromedial striatal neuroepithelium and subventricular zone

Subfornical organ

Stria terminalis

External capsule

Inferior frontal gyrus

Strionuclear glioepithelium

Stria medullaris

Choroid plexus

Bed nucleus of the stria terminalis

Subpial granular layer

Third ventricle

Globus pallidus (external segment)

Putamen

Insular gyrus

Lateral migratory stream (infiltrates claustrum and endopiriform nucleus)

BASAL GANGLIA

Fornix

Globus pallidus (internal segment)

Lateral fissure

Claustrum

Lateral migratory stream (extends toward olfactory tubercle)

PREOPTIC AREA

Lateral preoptic area

Anterior commissure

Ventral striatum

Claustrum

Uncinate fasciculus

Preoptic glioepithelium/ependyma

Magnocellular preoptic nucleus

SUBSTANTIA INNOMINATA

Olfactory tubercle

Layer I

Endopiriform nucleus

Cortical plate

Endopiriform nucleus

Sexually dimorphic nucleus

Medial forebrain bundle?

Primary olfactory cortex (Piriform)

Middle temporal gyrus

Layers III-IV

Medial preoptic area

Periventricular preoptic nucleus

Layer II

Primary olfactory cortex

Suprachiasmatic nucleus

Optic chiasm

Lateral olfactory tract

Layer I

White matter

Temporal lobe

Subplate

Lateral migratory stream (extends into the temporal lobe)

Inferior temporal gyrus

Germinal and transitional structures in *italics*

PLATE 84A
CR 350 mm
GW 37
Y217-65
Coronal
Section 761

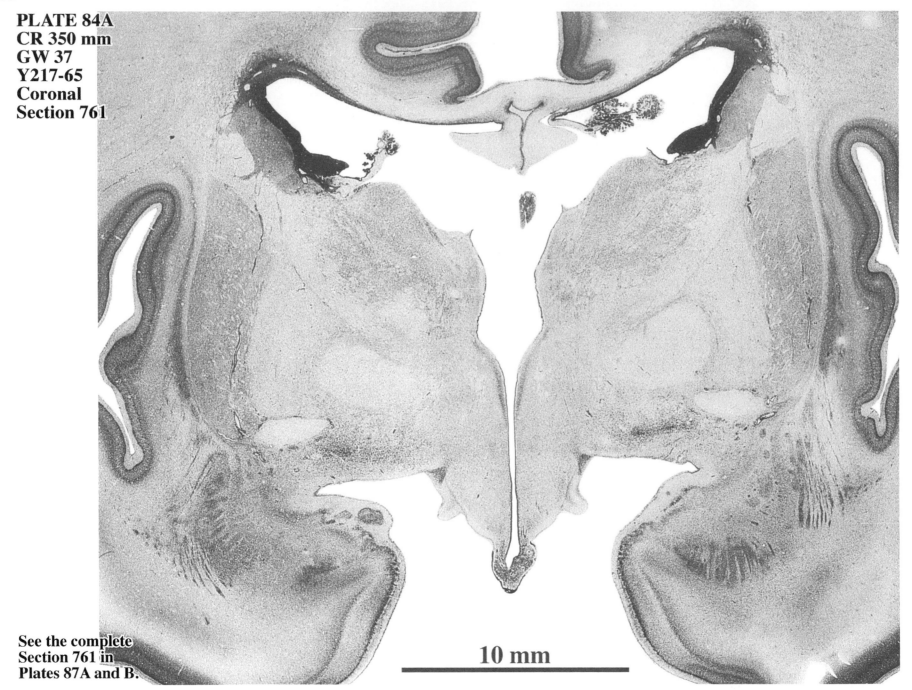

See the complete
Section 761 in
Plates 87A and B.

10 mm

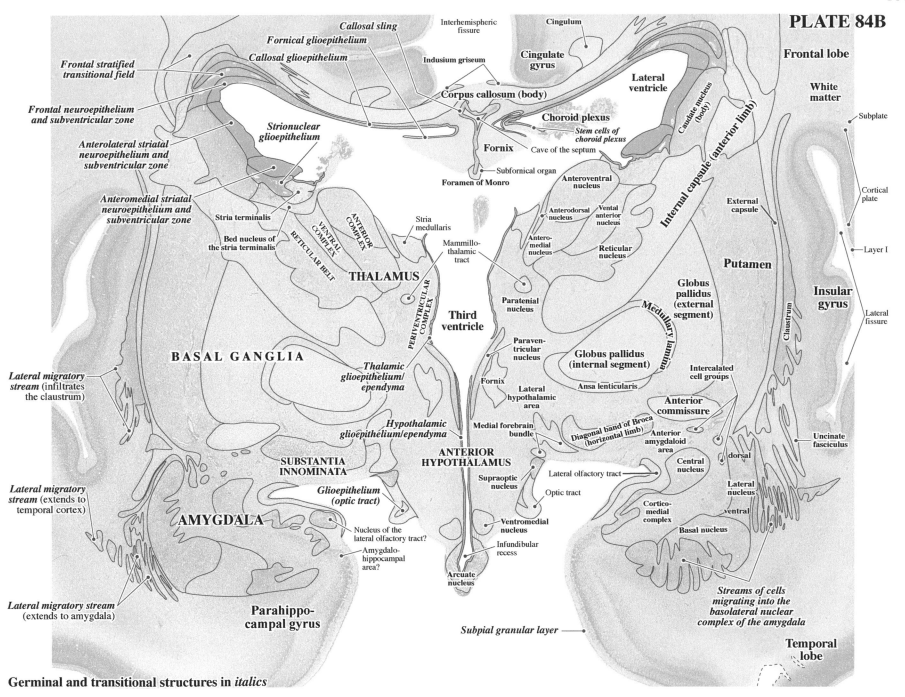

PLATE 84B

Callosal sling
Fornical glioepithelium
Callosal glioepithelium
Interhemispheric fissure
Cingulum
Cingulate gyrus
Indusium griseum
Frontal lobe

Frontal stratified transitional field

Corpus callosum (body)
Lateral ventricle
White matter
Subplate

Frontal neuroepithelium and subventricular zone

Choroid plexus
Caudate nucleus (body)
Internal capsule (anterior limb)
Cortical plate

Strionuclear glioepithelium
Stem cells of choroid plexus

Anterolateral striatal neuroepithelium and subventricular zone

Fornix
Cave of the septum

Anteroventral nucleus
External capsule

Anteromedial striatal neuroepithelium and subventricular zone

Subfornical organ
Anterodorsal nucleus
Ventral anterior nucleus
Layer I

Stria terminalis
Foramen of Monro
Antero-medial nucleus
Reticular nucleus

Bed nucleus of the stria terminalis
ANTERIOR COMPLEX
Stria medullaris
Mammillo-thalamic tract
Putamen

VENTRAL COMPLEX
RETICULAR BELT
Insular gyrus

THALAMUS
Globus pallidus (external segment)

Paratenial nucleus
Medullary lamina
Claustrum
Lateral fissure

Lateral migratory stream (infiltrates the claustrum)

BASAL GANGLIA
PERIVENTRICULAR COMPLEX
Third ventricle
Paraventricular nucleus
Globus pallidus (internal segment)

Thalamic glioepithelium/ ependyma
Fornix
Lateral hypothalamic area
Ansa lenticularis
Intercalated cell groups

Hypothalamic glioepithelium/ependyma
Medial forebrain bundle
Anterior commissure

Diagonal band of Broca (horizontal limb)
Anterior amygdaloid area
dorsal
Uncinate fasciculus

SUBSTANTIA INNOMINATA
ANTERIOR HYPOTHALAMUS
Central nucleus

Glioepithelium (optic tract)
Supraoptic nucleus
Lateral olfactory tract
Lateral nucleus

Lateral migratory stream (extends to temporal cortex)
Optic tract
Cortico-medial complex
ventral

AMYGDALA
Nucleus of the lateral olfactory tract?
Ventromedial nucleus
Basal nucleus

Amygdalo-hippocampal area?
Infundibular recess

Lateral migratory stream (extends to amygdala)
Arcuate nucleus
Streams of cells migrating into the basolateral nuclear complex of the amygdala

Parahippo-campal gyrus
Subpial granular layer
Temporal lobe

Germinal and transitional structures in *italics*

PLATE 85A
CR 350 mm
GW 37
Y217-65
Coronal
Section 831

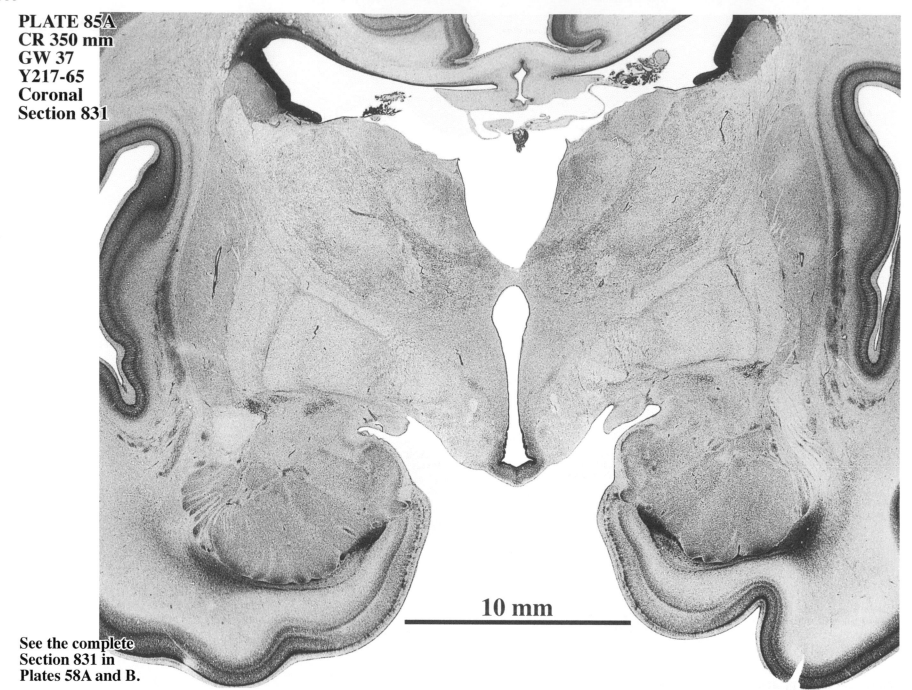

10 mm

See the complete
Section 831 in
Plates 58A and B.

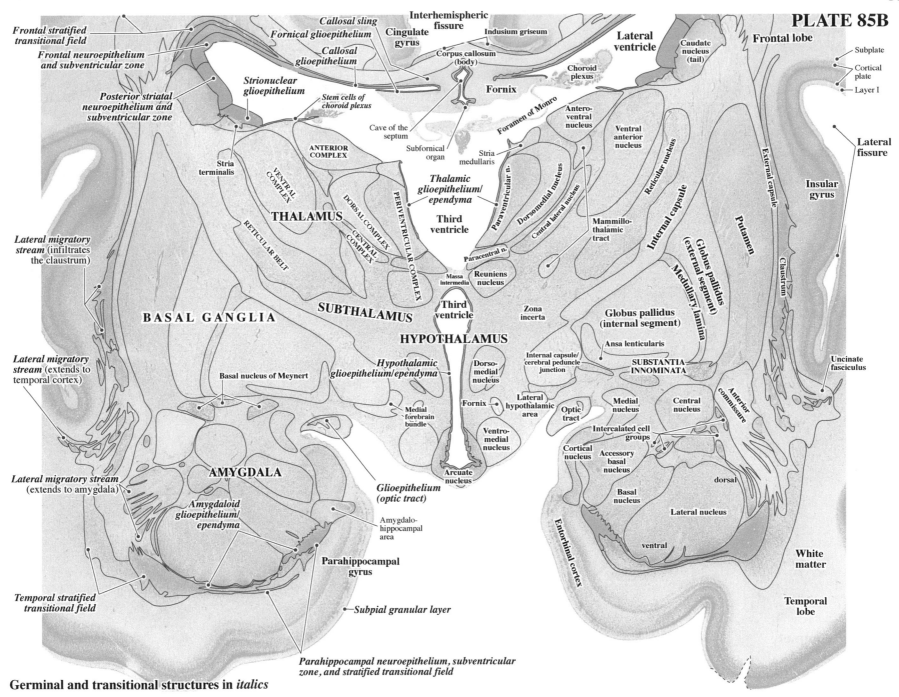

169

PLATE 85B

Frontal stratified transitional field

Frontal neuroepithelium and subventricular zone

Posterior striatal neuroepithelium and subventricular zone

Callosal sling

Fornical glioepithelium

Cingulate gyrus

Interhemispheric fissure

Indusium griseum

Lateral ventricle

Frontal lobe

Caudate nucleus (tail)

Corpus callosum (body)

Choroid plexus

Subplate

Cortical plate

Layer I

Callosal glioepithelium

Strionuclear glioepithelium

Stem cells of choroid plexus

Fornix

Foramen of Monro

Antero-ventral nucleus

Ventral anterior nucleus

Lateral fissure

Stria terminalis

ANTERIOR COMPLEX

Cave of the septum

Subfornical organ

Stria medullaris

Paraventricular n.

External capsule

Reticular nucleus

Insular gyrus

VENTRAL COMPLEX

THALAMUS

Thalamic glioepithelium/ ependyma

Third ventricle

Dorsomedial nucleus

Central lateral nucleus

Internal capsule

Putamen

DORSAL COMPLEX

CENTRAL COMPLEX

PERIVENTRICULAR COMPLEX

Mammillo-thalamic tract

Globus pallidus (external segment)

Medullary lamina

Claustrum

RETICULAR BELT

Massa intermedia

Paracentral n.

Reuniens nucleus

Lateral migratory stream (infiltrates the claustrum)

BASAL GANGLIA

SUBTHALAMUS

Third ventricle

Zona incerta

Globus pallidus (internal segment)

HYPOTHALAMUS

Ansa lenticularis

Lateral migratory stream (extends to temporal cortex)

Basal nucleus of Meynert

Hypothalamic glioepithelium/ependyma

Dorso-medial nucleus

Internal capsule/ cerebral peduncle junction

SUBSTANTIA INNOMINATA

Uncinate fasciculus

Fornix

Lateral hypothalamic area

Medial forebrain bundle

Optic tract

Medial nucleus

Central nucleus

Anterior commissure

Lateral migratory stream (extends to amygdala)

AMYGDALA

Amygdaloid glioepithelium/ ependyma

Ventro-medial nucleus

Glioepithelium (optic tract)

Intercalated cell groups

dorsal

Cortical nucleus

Accessory basal nucleus

Arcuate nucleus

Amygdalo-hippocampal area

Basal nucleus

Lateral nucleus

White matter

Parahippocampal gyrus

Entorhinal cortex

ventral

Temporal stratified transitional field

—Subpial granular layer

Temporal lobe

Parahippocampal neuroepithelium, subventricular zone, and stratified transitional field

Germinal and transitional structures in *italics*

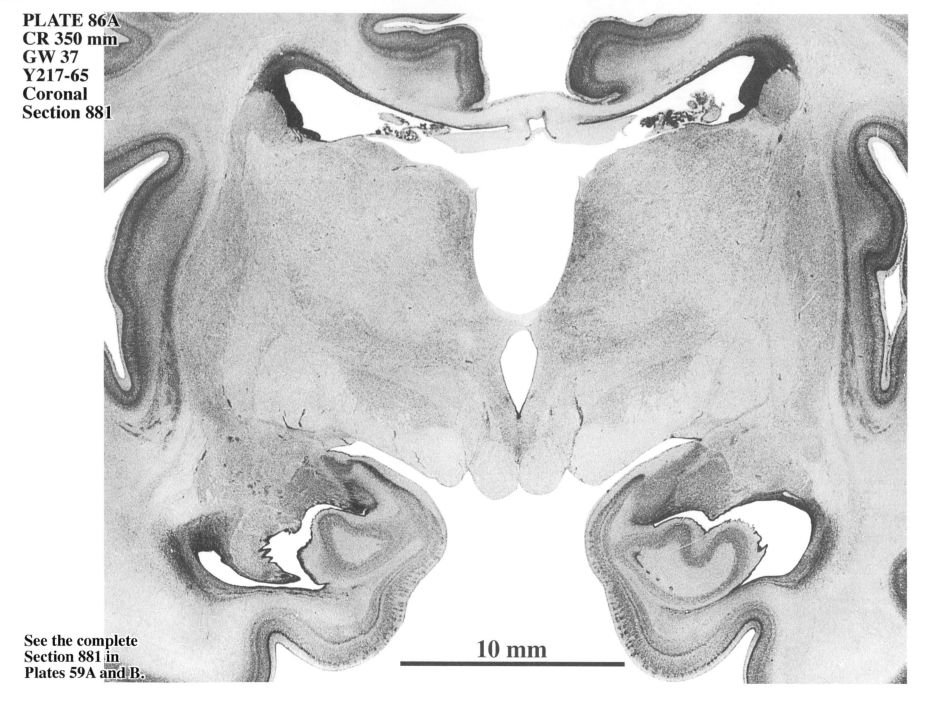

PLATE 86A
CR 350 mm
GW 37
Y217-65
Coronal
Section 881

See the complete
Section 881 in
Plates 59A and B.

10 mm

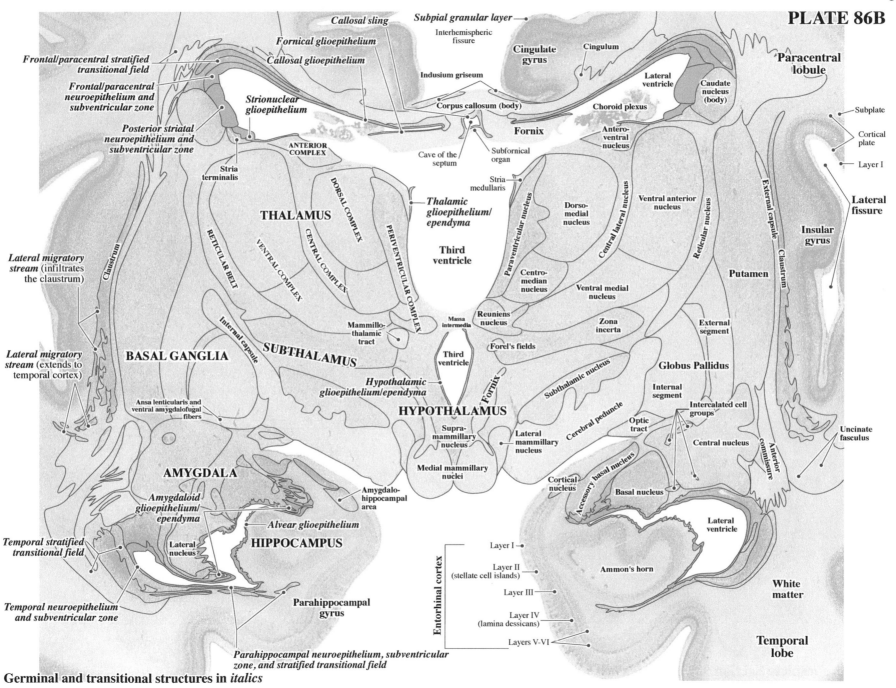

Frontal/paracentral stratified transitional field

Frontal/paracentral neuroepithelium and subventricular zone

Posterior striatal neuroepithelium and subventricular zone

Callosal sling

Fornical glioepithelium

Callosal glioepithelium

Strionuclear glioepithelium

Subpial granular layer

Interhemispheric fissure

Indusium griseum

Corpus callosum (body)

Cingulate gyrus

Cingulum

Lateral ventricle

Caudate nucleus (body)

Choroid plexus

Fornix

Antero-ventral nucleus

Paracentral lobule

Subplate

Cortical plate

Layer I

Stria terminalis

ANTERIOR COMPLEX

Cave of the septum

Subfornical organ

Stria medullaris

Thalamic glioepithelium/ ependyma

DORSAL COMPLEX

THALAMUS

RETICULAR BELT

VENTRAL COMPLEX

CENTRAL COMPLEX

PERIVENTRICULAR COMPLEX

Third ventricle

Paraventricular nucleus

Dorso-medial nucleus

Ventral anterior nucleus

Central lateral nucleus

Reticular nucleus

External capsule

Claustrum

Insular gyrus

Lateral fissure

Lateral migratory stream (infiltrates the claustrum)

Claustrum

Lateral migratory stream (extends to temporal cortex)

Internal capsule

BASAL GANGLIA

SUBTHALAMUS

Mammillo-thalamic tract

Massa intermedia

Reuniens nucleus

Centro-median nucleus

Ventral medial nucleus

Zona incerta

Forel's fields

Third ventricle

Putamen

External segment

Globus Pallidus

Internal segment

Intercalated cell groups

Ansa lenticularis and ventral amygdalofugal fibers

Hypothalamic glioepithelium/ependyma

HYPOTHALAMUS

Fornix

Subthalamic nucleus

Cerebral peduncle

Optic tract

Central nucleus

Anterior commissure

Uncinate fasciculus

Supra-mammillary nucleus

Lateral mammillary nucleus

AMYGDALA

Amygdaloid glioepithelium/ ependyma

Lateral nucleus

Alvear glioepithelium

HIPPOCAMPUS

Amygdalo-hippocampal area

Medial mammillary nuclei

Cortical nucleus

Accessory basal nucleus

Basal nucleus

Lateral ventricle

Temporal stratified transitional field

Temporal neuroepithelium and subventricular zone

Parahippocampal gyrus

Entorhinal cortex

Layer I

Layer II (stellate cell islands)

Layer III

Layer IV (lamina dessicans)

Layers V-VI

Ammon's horn

White matter

Temporal lobe

Parahippocampal neuroepithelium, subventricular zone, and stratified transitional field

Germinal and transitional structures in *italics*

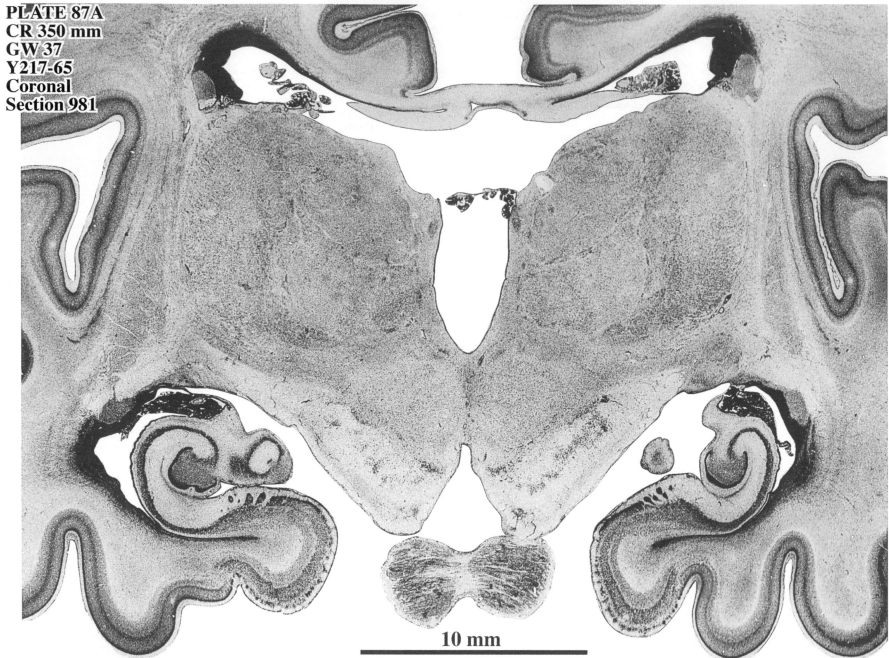

PLATE 87A
CR 350 mm
GW 37
Y217-65
Coronal
Section 981

10 mm

See the complete Section 981 in Plates 60A and B.

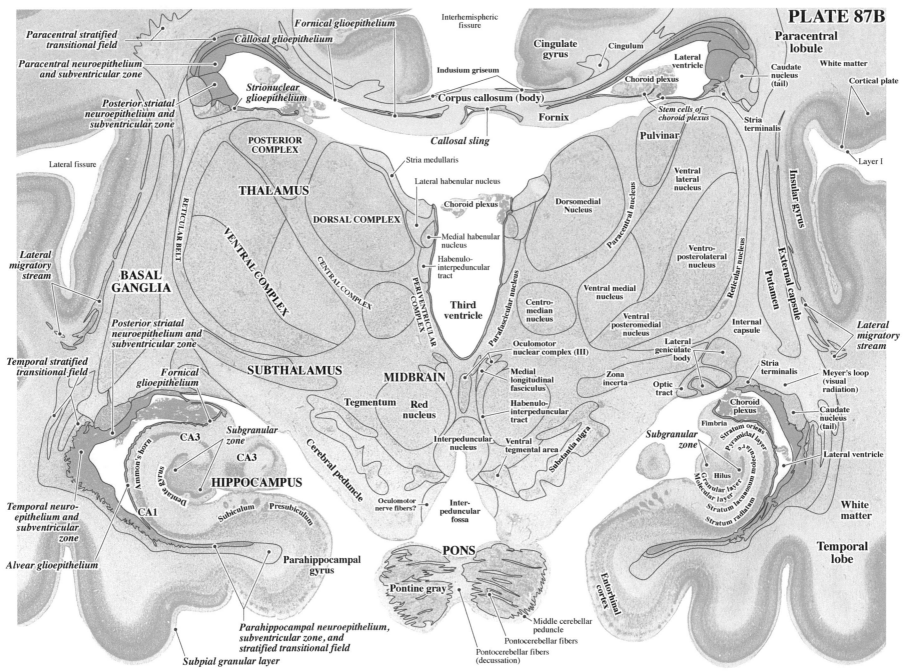

PLATE 87B

Interhemispheric fissure

Paracentral stratified transitional field

Fornical glioepithelium

Callosal glioepithelium

Cingulate gyrus

Cingulum

Paracentral lobule

Paracentral neuroepithelium and subventricular zone

Lateral ventricle

Caudate nucleus (tail)

White matter

Choroid plexus

Cortical plate

Posterior striatal neuroepithelium and subventricular zone

Indusium griseum

Strionuclear glioepithelium

Corpus callosum (body)

Pulvinar

Stem cells of choroid plexus

Fornix

Stria terminalis

Lateral fissure

POSTERIOR COMPLEX

Callosal sling

Stria medullaris

Ventral lateral nucleus

Layer I

Lateral habenular nucleus

Dorsomedial Nucleus

THALAMUS

Choroid plexus

RETICULAR BELT

DORSAL COMPLEX

Medial habenular nucleus

Paracentral nucleus

Ventro-posterolateral nucleus

Reticular nucleus

Lateral migratory stream

VENTRAL COMPLEX

Habenulo-interpeduncular tract

External capsule

Putamen

BASAL GANGLIA

CENTRAL COMPLEX

Centro-median nucleus

Ventral medial nucleus

Parafascicular nucleus

PERIVENTRICULAR COMPLEX

Third ventricle

Ventral posteromedial nucleus

Internal capsule

Posterior striatal neuroepithelium and subventricular zone

Oculomotor nuclear complex (III)

Lateral geniculate body

Stria terminalis

Lateral migratory stream

Temporal stratified transitional field

SUBTHALAMUS

MIDBRAIN

Medial longitudinal fasciculus

Zona incerta

Optic tract

Meyer's loop (visual radiation)

Fornical glioepithelium

Tegmentum

Red nucleus

Habenulo-interpeduncular tract

Choroid plexus

Caudate nucleus (tail)

Temporal neuro-epithelium and subventricular zone

CA3

Subgranular zone

Cerebral peduncle

Fimbria

Stratum oriens

Subgranular zone

Pyramidal layer

Ammon's horn

CA3

HIPPOCAMPUS

Interpeduncular nucleus

Ventral tegmental area

Substantia nigra

Lateral ventricle

Dentate gyrus

CA1

Subiculum

Presubiculum

Oculomotor nerve fibers?

Inter-peduncular fossa

Hilus

Granular layer

Molecular layer

Stratum lacunosum

Stratum radiatum

Alvear glioepithelium

Parahippocampal gyrus

White matter

Entorhinal cortex

Temporal lobe

Parahippocampal neuroepithelium, subventricular zone, and stratified transitional field

PONS

Pontine gray

Middle cerebellar peduncle

Subpial granular layer

Pontocerebellar fibers

Pontocerebellar fibers (decussation)

Germinal and transitional structures in *italics*

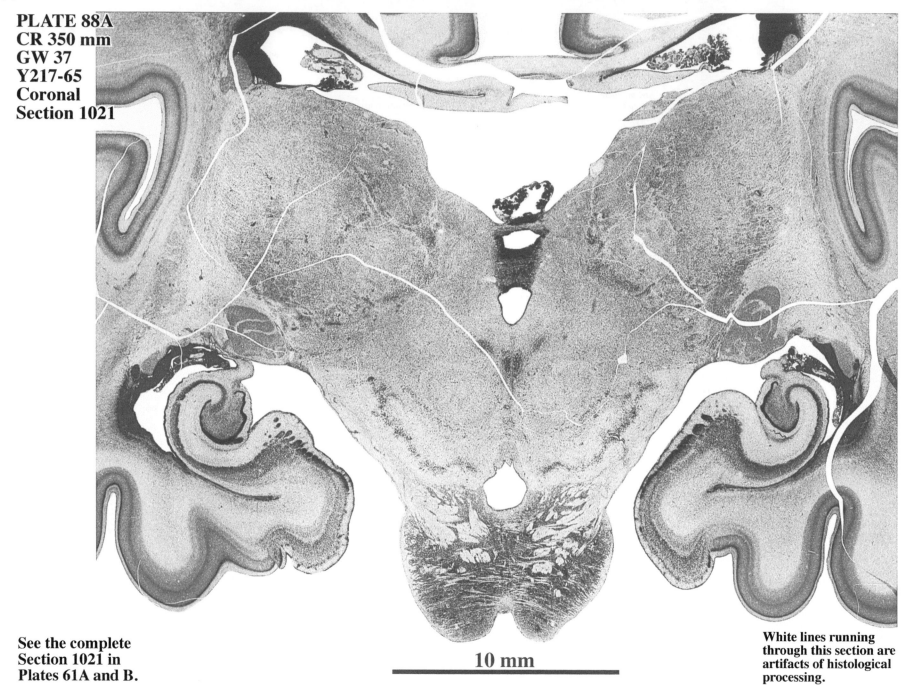

PLATE 88A
CR 350 mm
GW 37
Y217-65
Coronal
Section 1021

See the complete
Section 1021 in
Plates 61A and B.

10 mm

White lines running
through this section are
artifacts of histological
processing.

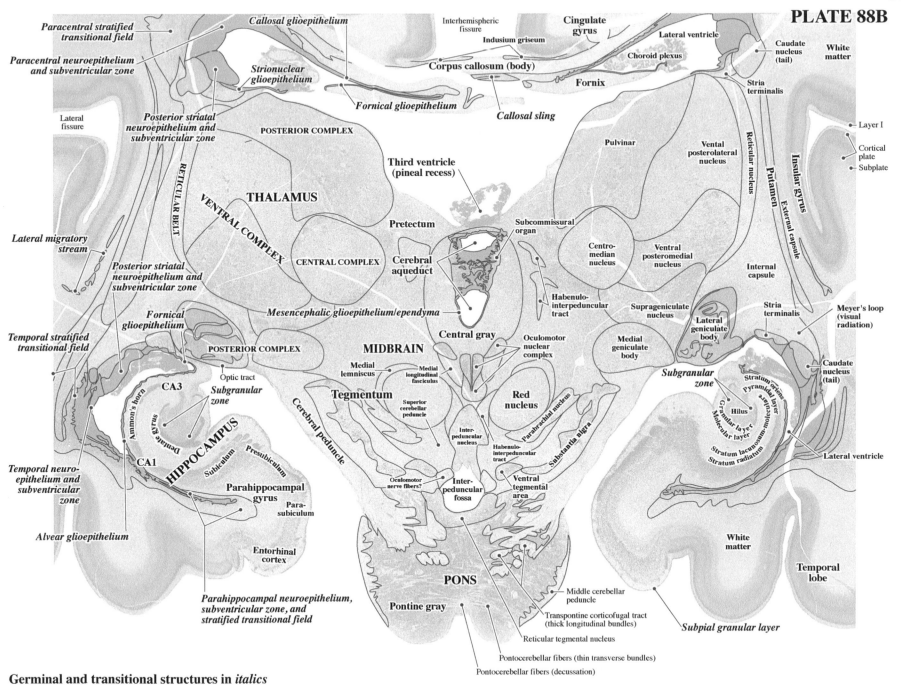

Paracentral stratified transitional field

Paracentral neuroepithelium and subventricular zone

Callosal glioepithelium

Interhemispheric fissure

Cingulate gyrus

Lateral ventricle

Caudate nucleus (tail)

White matter

Indusium griseum

Choroid plexus

Corpus callosum (body)

Strionuclear glioepithelium

Stria terminalis

Fornix

Fornical glioepithelium

Callosal sling

Lateral fissure

Posterior striatal neuroepithelium and subventricular zone

POSTERIOR COMPLEX

Pulvinar

Vental posterolateral nucleus

Reticular nucleus

Insular gyrus

Layer I

RETICULAR BELT

VENTRAL COMPLEX

THALAMUS

Third ventricle (pineal recess)

Subcommissural organ

Cortical plate

Subplate

Putamen

External capsule

Lateral migratory stream

Pretectum

Centro-median nucleus

Ventral posteromedial nucleus

Posterior striatal neuroepithelium and subventricular zone

CENTRAL COMPLEX

Cerebral aqueduct

Internal capsule

Habenulo-interpeduncular tract

Suprageniculate nucleus

Stria terminalis

Fornical glioepithelium

Mesencephalic glioepithelium/ependyma

Central gray

Oculomotor nuclear complex

Medial geniculate body

Lateral geniculate body

Meyer's loop (visual radiation)

Temporal stratified transitional field

POSTERIOR COMPLEX

MIDBRAIN

Subgranular zone

Caudate nucleus (tail)

Optic tract

Medial lemniscus

Medial longitudinal fasciculus

Stratum oriens

Pyramidal layer

CA3

Subgranular zone

Tegmentum

Superior cerebellar peduncle

Red nucleus

Parabrachial nucleus

Granular layer

Molecular layer

Hilus

Cerebral peduncle

Temporal neuro-epithelium and subventricular zone

Ammon's horn

Dentate gyrus

Inter-peduncular nucleus

Habenulo-interpeduncular tract

Substantia nigra

Stratum lacunosum-molecular

Lateral ventricle

CA1

HIPPOCAMPUS

Subiculum

Presubiculum

Oculomotor nerve fibers?

Inter-peduncular fossa

Ventral tegmental area

Stratum radiatum

Alvear glioepithelium

Parahippocampal gyrus

Para-subiculum

Entorhinal cortex

Parahippocampal neuroepithelium, subventricular zone, and stratified transitional field

PONS

Pontine gray

Middle cerebellar peduncle

Transpontine corticofugal tract (thick longitudinal bundles)

White matter

Temporal lobe

Subpial granular layer

Reticular tegmental nucleus

Pontocerebellar fibers (thin transverse bundles)

Pontocerebellar fibers (decussation)

Germinal and transitional structures in *italics*

PLATE 89A
CR 350 mm
GW 37
Y217-65
Coronal
Section 1081

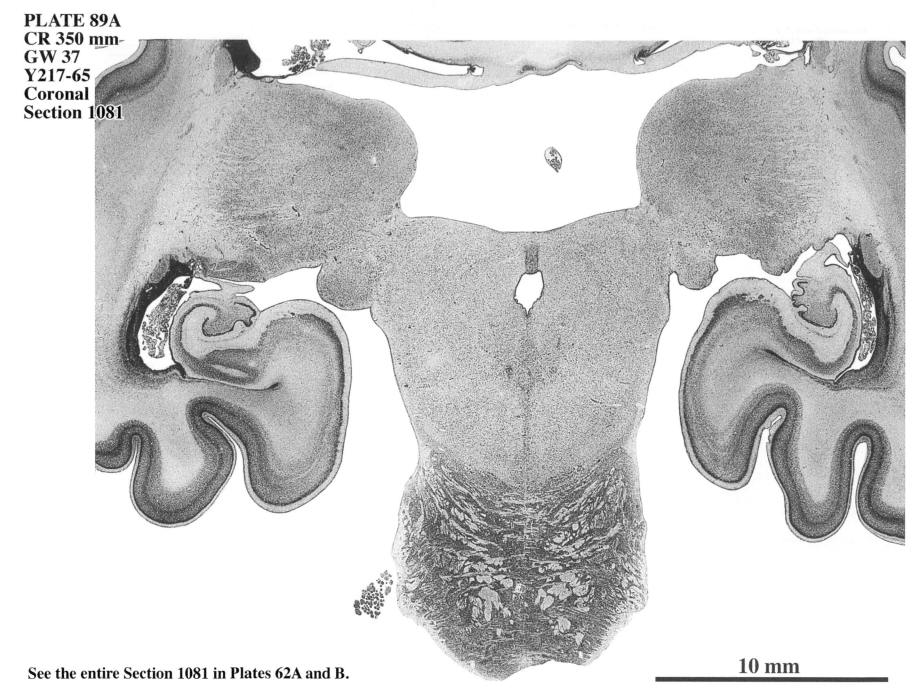

See the entire Section 1081 in Plates 62A and B.

10 mm

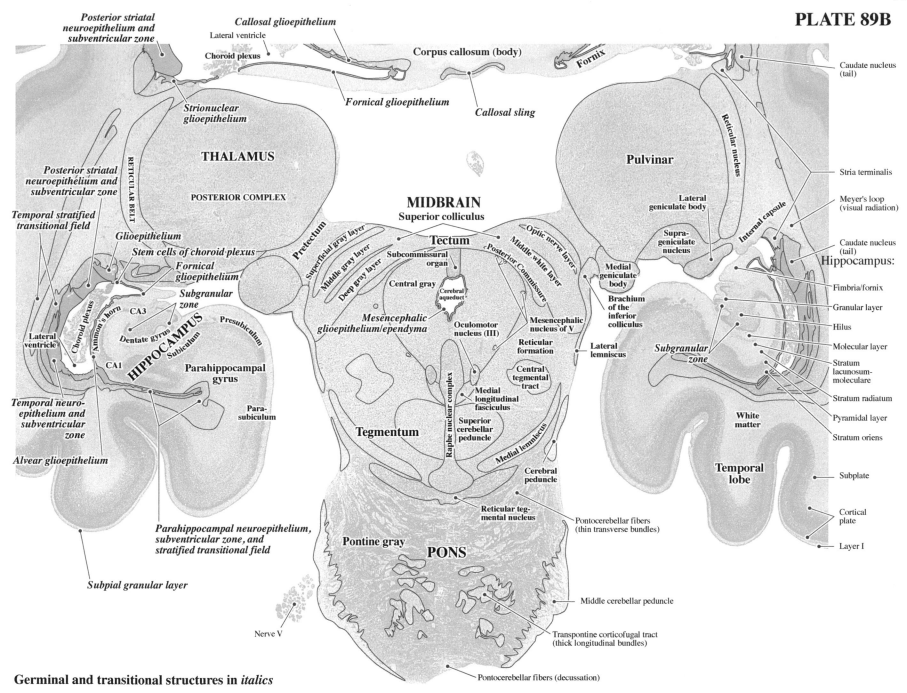

Germinal and transitional structures in *italics*

PLATE 90A
CR 350 mm, GW 37, Y217-65, Coronal, Section 1141

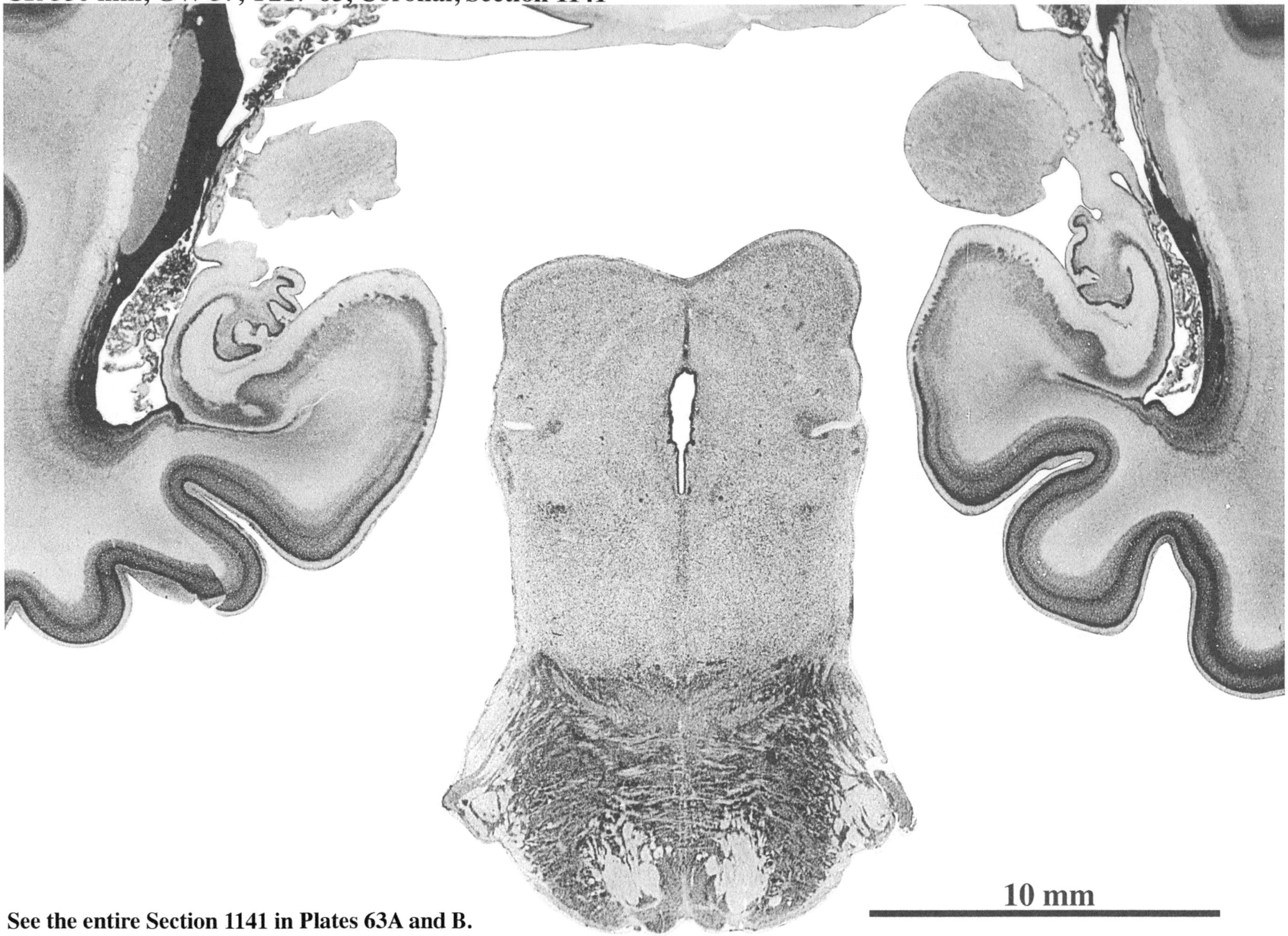

10 mm

See the entire Section 1141 in Plates 63A and B.

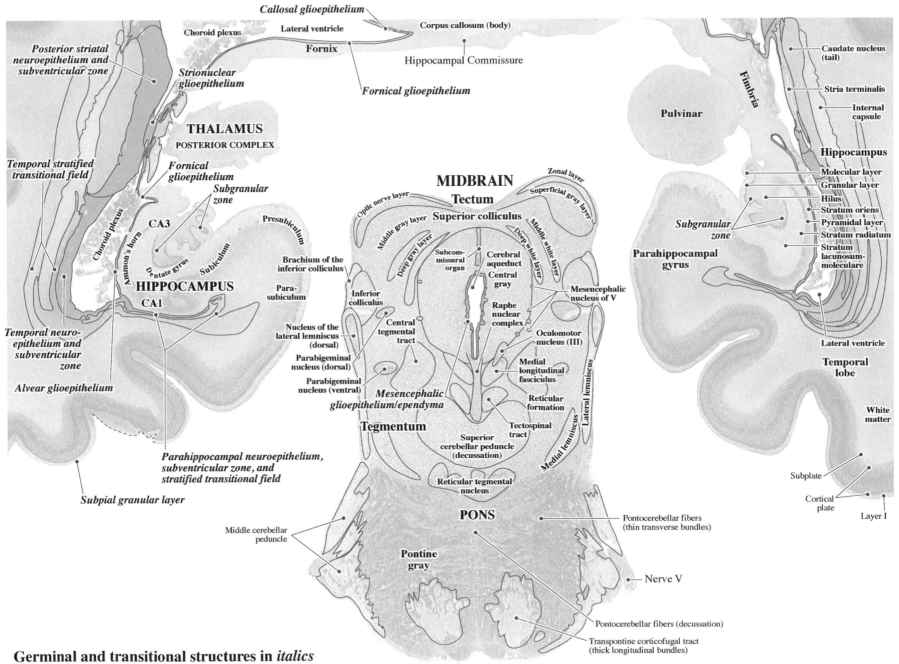

Callosal glioepithelium

Choroid plexus

Lateral ventricle

Corpus callosum (body)

Fornix

Posterior striatal neuroepithelium and subventricular zone

Hippocampal Commissure

Strionuclear glioepithelium

Fornical glioepithelium

Caudate nucleus (tail)

Stria terminalis

Fimbria

Internal capsule

Pulvinar

THALAMUS

POSTERIOR COMPLEX

Fornical glioepithelium

Temporal stratified transitional field

Subgranular zone

MIDBRAIN

Tectum

Zonal layer

Superficial gray layer

Hippocampus

Molecular layer

Granular layer

Hilus

Choroid plexus

CA3

Presubiculum

Optic nerve layer

Superior colliculus

Middle gray layer

Middle white layer

Deep white layer

Subgranular zone

Stratum oriens

Pyramidal layer

Stratum radiatum

Stratum lacunosum-moleculare

Dentate gyrus

Subiculum

Deep gray layer

Subcommissural organ

Cerebral aqueduct

Central gray

Mesencephalic nucleus of V

Parahippocampal gyrus

Ammon's horn

HIPPOCAMPUS

CA1

Para-subiculum

Brachium of the inferior colliculus

Inferior colliculus

Raphe nuclear complex

Oculomotor nucleus (III)

Lateral ventricle

Temporal neuro-epithelium and subventricular zone

Nucleus of the lateral lemniscus (dorsal)

Central tegmental tract

Medial longitudinal fasciculus

Temporal lobe

Alvear glioepithelium

Parabigeminal nucleus (dorsal)

Parabigeminal nucleus (ventral)

Lateral lemniscus

Reticular formation

Mesencephalic glioepithelium/ependyma

Tectospinal tract

White matter

Tegmentum

Superior cerebellar peduncle (decussation)

Medial lemniscus

Parahippocampal neuroepithelium, subventricular zone, and stratified transitional field

Reticular tegmental nucleus

Subplate

Subpial granular layer

PONS

Pontocerebellar fibers (thin transverse bundles)

Cortical plate

Layer I

Middle cerebellar peduncle

Pontine gray

Nerve V

Pontocerebellar fibers (decussation)

Transpontine corticofugal tract (thick longitudinal bundles)

Germinal and transitional structures in *italics*

PLATE 91A
CR 350 mm, GW 37, Y217-65, Coronal, Section 1181

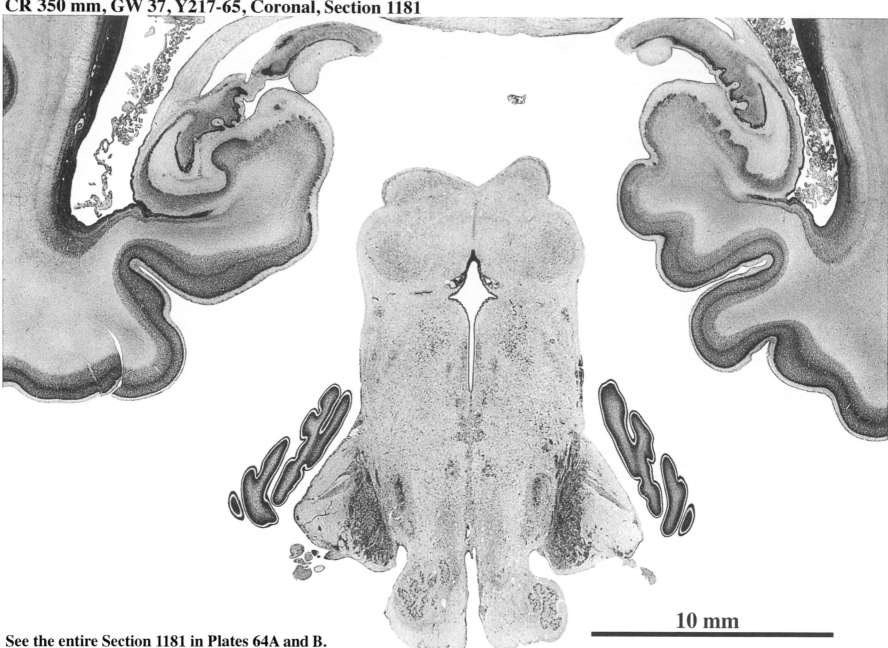

10 mm

See the entire Section 1181 in Plates 64A and B.

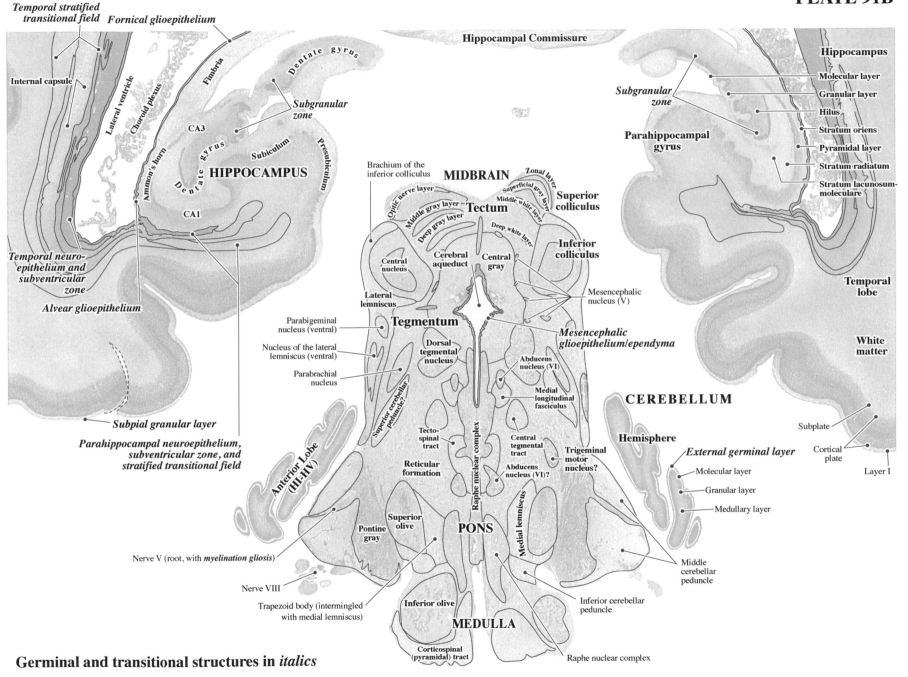

Temporal stratified transitional field

Fornical glioepithelium

Internal capsule

Lateral ventricle

Choroid plexus

Fimbria

Dentate gyrus

Subgranular zone

CA3

Dentate gyrus

HIPPOCAMPUS

Ammon's horn

Subiculum

Presubiculum

CA1

Temporal neuro-epithelium and subventricular zone

Alvear glioepithelium

Subpial granular layer

Parahippocampal neuroepithelium, subventricular zone, and stratified transitional field

Hippocampal Commissure

Brachium of the inferior colliculus

MIDBRAIN

Zonal layer

Superficial gray layer

Middle white layer

Superior colliculus

Optic nerve layer

Middle gray layer

Deep gray layer

Tectum

Deep white layer

Inferior colliculus

Central nucleus

Cerebral aqueduct

Central gray

Lateral lemniscus

Mesencephalic nucleus (V)

Parabigeminal nucleus (ventral)

Tegmentum

Mesencephalic glioepithelium/ependyma

Nucleus of the lateral lemniscus (ventral)

Dorsal tegmental nucleus

Abducens nucleus (VI)

Parabrachial nucleus

Medial longitudinal fasciculus

Superior cerebellar peduncle?

Tecto-spinal tract

Raphe nuclear complex

Central tegmental tract

Trigeminal motor nucleus?

Reticular formation

Abducens nucleus (VI)?

Anterior Lobe (HI-HV)

Superior olive

Pontine gray

PONS

Medial lemniscus

Nerve V (root, with *myelination gliosis*)

Nerve VIII

Trapezoid body (intermingled with medial lemniscus)

MEDULLA

Inferior olive

Corticospinal (pyramidal) tract

Raphe nuclear complex

CEREBELLUM

Hippocampus

Molecular layer

Granular layer

Hilus

Stratum oriens

Pyramidal layer

Stratum radiatum

Stratum lacunosum-moleculare

Subgranular zone

Parahippocampal gyrus

Temporal lobe

White matter

Hemisphere

External germinal layer

Molecular layer

Granular layer

Medullary layer

Subplate

Cortical plate

Layer I

Middle cerebellar peduncle

Inferior cerebellar peduncle

Germinal and transitional structures in *italics*

PLATE 92A
CR 350 mm
GW 37
Y217-65
Coronal
Section 1221

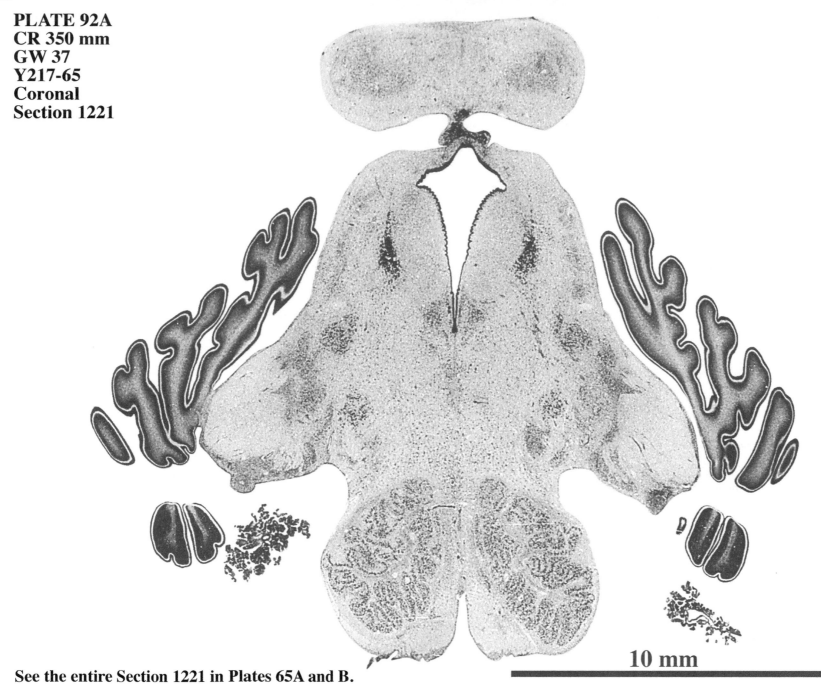

10 mm

See the entire Section 1221 in Plates 65A and B.

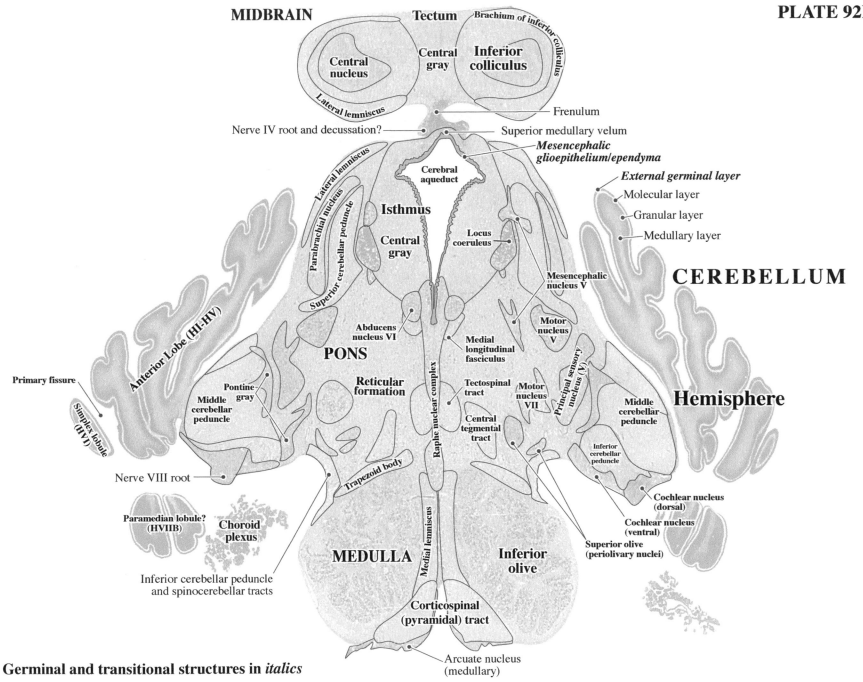

MIDBRAIN   Tectum   Brachium of inferior colliculus

Central nucleus   Central gray   Inferior colliculus

Lateral lemniscus

Frenulum

Nerve IV root and decussation? — Superior medullary velum

Mesencephalic glioepithelium/ependyma

Cerebral aqueduct

External germinal layer

Molecular layer

Granular layer

Medullary layer

Lateral lemniscus

Parabrachial nucleus

Superior cerebellar peduncle

Isthmus

Central gray

Locus coeruleus

CEREBELLUM

Mesencephalic nucleus V

Motor nucleus V

Abducens nucleus VI

Medial longitudinal fasciculus

PONS

Anterior Lobe (HI–HV)

Primary fissure

Reticular formation

Tectospinal tract

Motor nucleus VII

Principal sensory nucleus (V)

Middle cerebellar peduncle

Hemisphere

Simplex lobule (HVI)

Pontine gray

Middle cerebellar peduncle

Central tegmental tract

Inferior cerebellar peduncle

Nerve VIII root

Trapezoid body

Cochlear nucleus (dorsal)

Cochlear nucleus (ventral)

Paramedian lobule? (HVIIB)

Choroid plexus

Raphe nuclear complex

Medial lemniscus

MEDULLA

Inferior olive

Superior olive (periolivary nuclei)

Inferior cerebellar peduncle and spinocerebellar tracts

Corticospinal (pyramidal) tract

Arcuate nucleus (medullary)

Germinal and transitional structures in *italics*

PLATE 93A
CR 350 mm
GW 37
Y217-65
Coronal
Section 1311

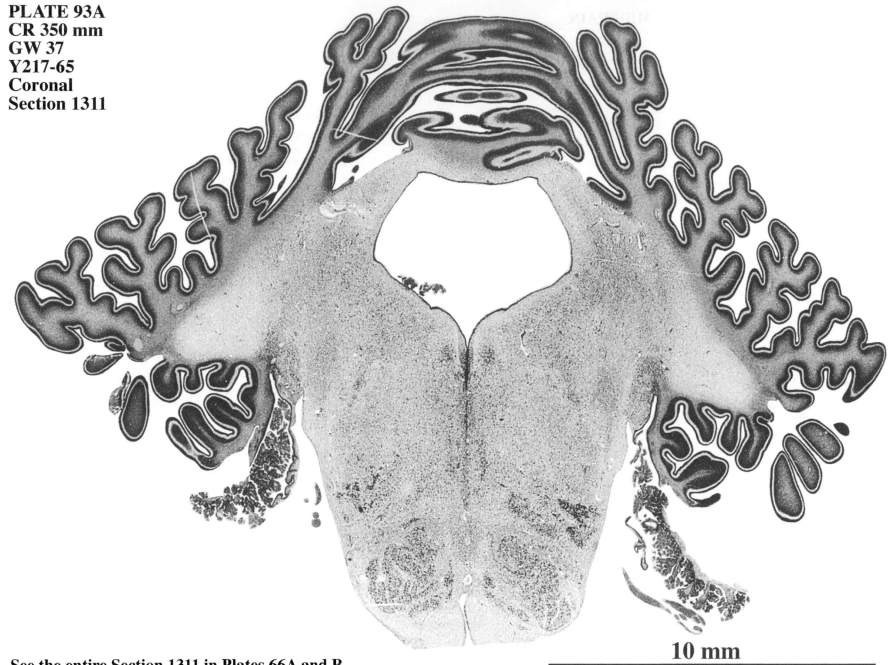

See the entire Section 1311 in Plates 66A and B.

10 mm

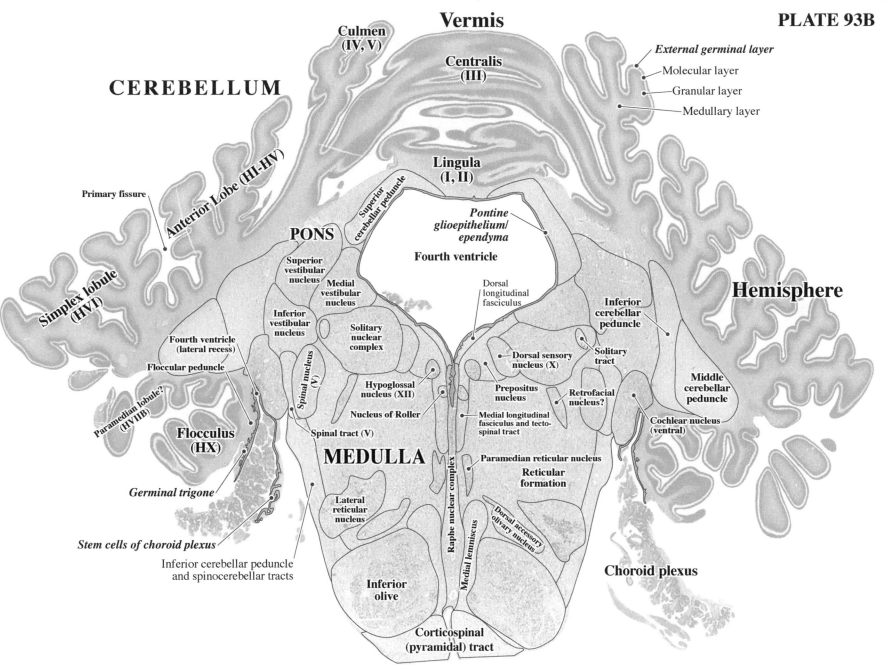

Vermis

Culmen (IV, V)

Centralis (III)

External germinal layer
Molecular layer
Granular layer
Medullary layer

Lingula (I, II)

CEREBELLUM

Anterior Lobe (HI-HV)

Primary fissure

Superior cerebellar peduncle

Pontine glioepithelium/ ependyma

PONS

Fourth ventricle

Simplex lobule (HVI)

Superior vestibular nucleus

Medial vestibular nucleus

Dorsal longitudinal fasciculus

Hemisphere

Inferior cerebellar peduncle

Inferior vestibular nucleus

Solitary nuclear complex

Dorsal sensory nucleus (X)

Solitary tract

Fourth ventricle (lateral recess)

Spinal nucleus (V)

Hypoglossal nucleus (XII)

Prepositus nucleus

Retrofacial nucleus?

Middle cerebellar peduncle

Floccular peduncle

Nucleus of Roller

Medial longitudinal fasciculus and tecto-spinal tract

Cochlear nucleus (ventral)

Paramedian lobule? (HVIIB)

Spinal tract (V)

Flocculus (HX)

MEDULLA

Paramedian reticular nucleus

Reticular formation

Germinal trigone

Lateral reticular nucleus

Raphe nuclear complex

Dorsal accessory olivary nucleus

Choroid plexus

Stem cells of choroid plexus

Inferior cerebellar peduncle and spinocerebellar tracts

Inferior olive

Medial lemniscus

Corticospinal (pyramidal) tract

Germinal and transitional structures in *italics*

PLATE 94A
CR 350 mm
GW 37
Y217-65
Coronal
Section 1371

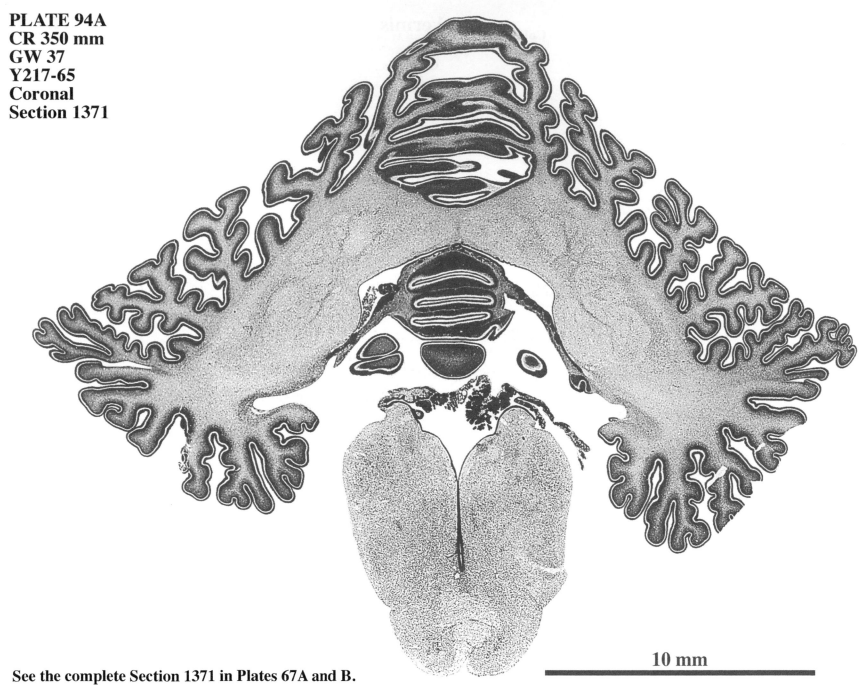

See the complete Section 1371 in Plates 67A and B.

10 mm

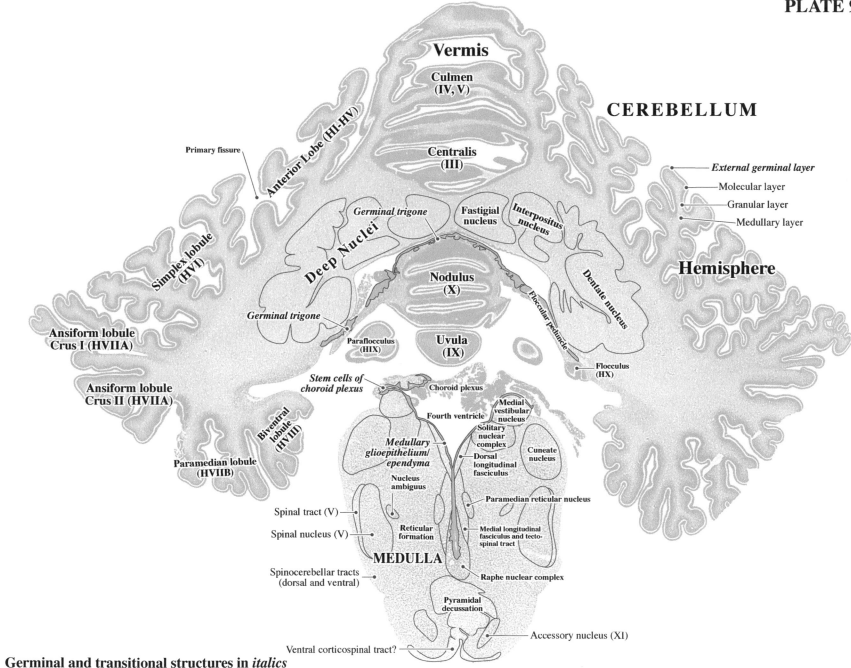

Germinal and transitional structures in *italics*

PLATE 95A
CR 350 mm
GW 37
Y217-65
Coronal
Section 1501

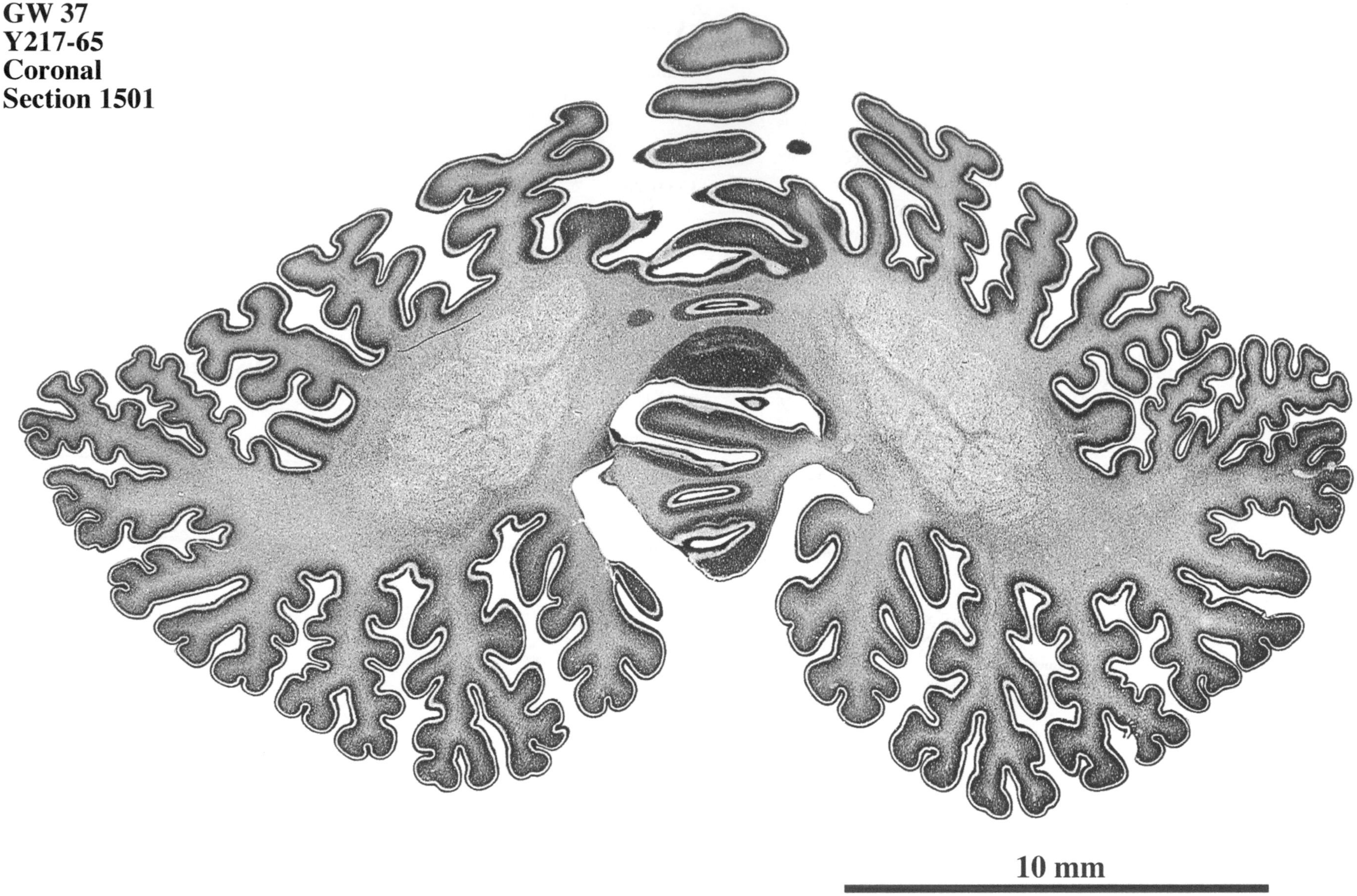

10 mm

See the complete Section 1501 in Plates 68A and B.

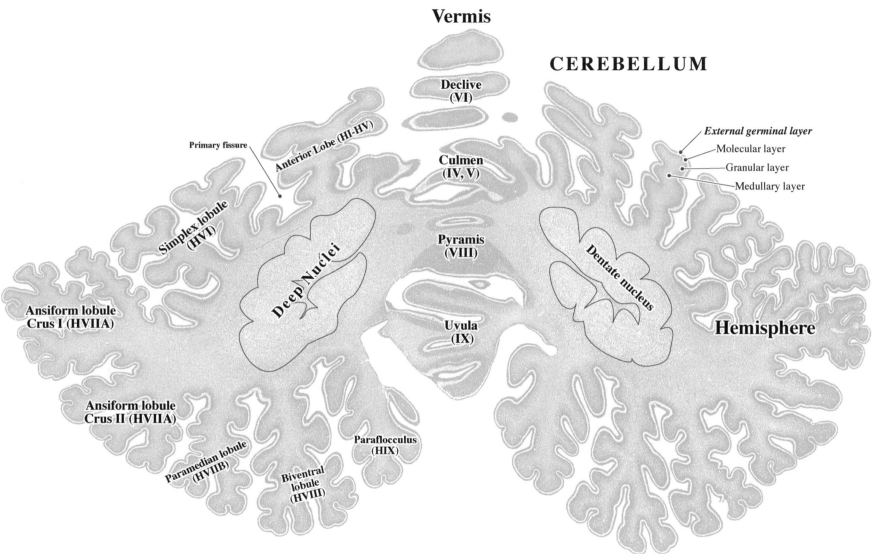

Vermis

CEREBELLUM

Declive
(VI)

External germinal layer

Molecular layer

Primary fissure

Anterior Lobe (HI-HV)

Culmen
(IV, V)

Granular layer

Medullary layer

Simplex lobule
(HVI)

Pyramis
(VIII)

Deep Nuclei

Dentate nucleus

Ansiform lobule
Crus I (HVIIA)

Uvula
(IX)

Hemisphere

Ansiform lobule
Crus II (HVIIA)

Paraflocculus
(HIX)

Paramedian lobule
(HVIIB)

Biventral
lobule
(HVIII)

Germinal and transitional structures in *italics*

PLATE 96A
CR 350 mm
GW 37
Y217-65
Coronal
Section 1611

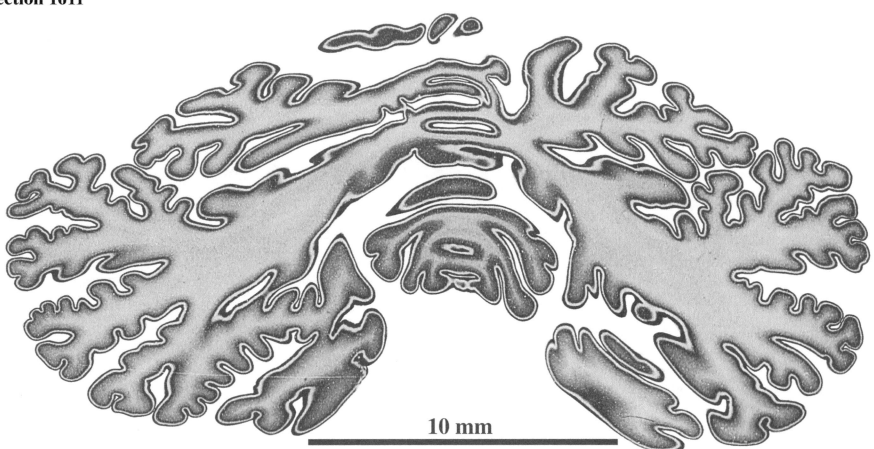

10 mm

See the complete Section 1611 in Plates 69A and B.

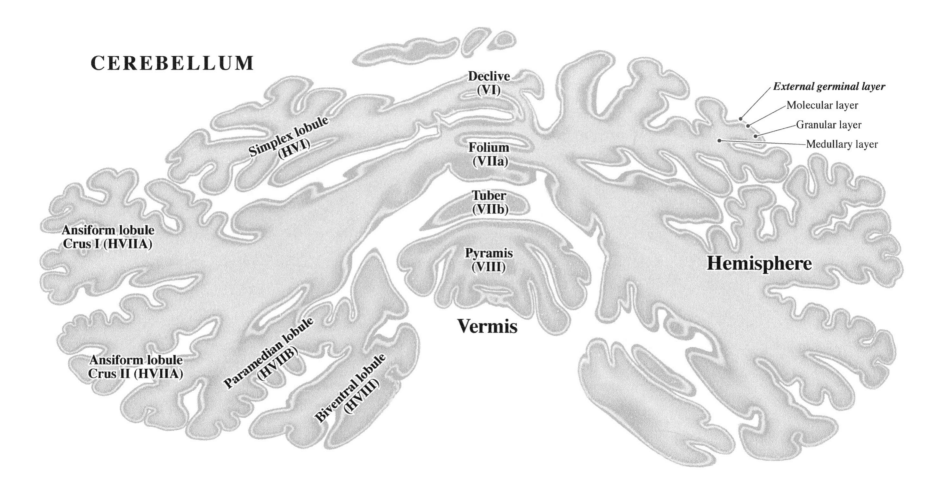

CEREBELLUM

Declive
(VI)

External germinal layer

Molecular layer

Granular layer

Medullary layer

Simplex lobule
(HVI)

Folium
(VIIa)

Tuber
(VIIb)

Ansiform lobule
Crus I (HVIIA)

Pyramis
(VIII)

Hemisphere

Vermis

Ansiform lobule
Crus II (HVIIA)

Paramedian lobule
(HVIIB)

Biventral lobule
(HVIII)

Germinal and transitional structures in *italics*

PLATE 97A
CR 350 mm
GW 37
Y217-65
Coronal
Section 1711

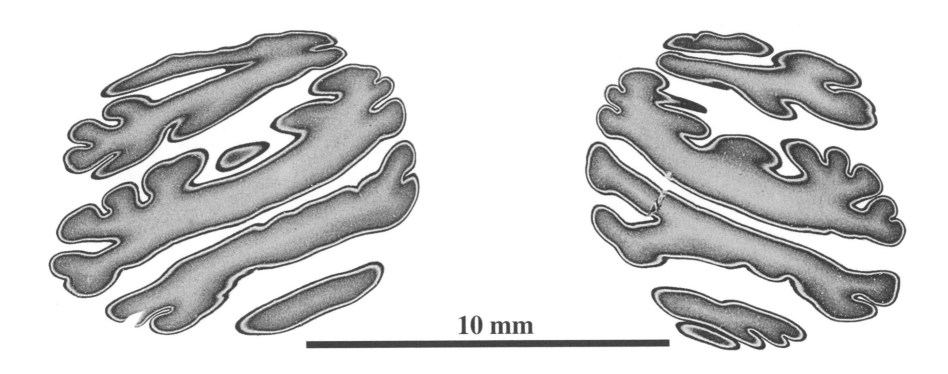

10 mm

See the complete Section 1711 in Plates 70A and B.

CEREBELLUM

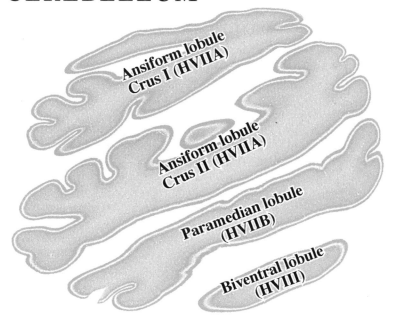

Ansiform lobule
Crus I (HVIIA)

Ansiform lobule
Crus II (HVIIA)

Paramedian lobule
(HVIIB)

Biventral lobule
(HVIII)

Hemisphere

T - #0910 - 101024 - C51 - 210/279/9 - PB - 9781032228808 - Gloss Lamination